ROUTLEDGE LIBRARY EDITIONS:
ENVIRONMENTAL AND NATURAL
RESOURCE ECONOMICS

Volume 13

# MINING AND DEVELOPMENT

# MINING AND DEVELOPMENT

## Foreign-Financed Mines in Australia, Ireland, Papua New Guinea and Zambia

CIARAN O'FAIRCHEALLAIGH

Routledge
Taylor & Francis Group

LONDON AND NEW YORK

First published in 1984 by Croom Helm Ltd

This edition first published in 2018
by Routledge
2 Park Square, Milton Park, Abingdon, Oxon OX14 4RN

and by Routledge
711 Third Avenue, New York, NY 10017

*Routledge is an imprint of the Taylor & Francis Group, an informa business*

*British Library Cataloguing in Publication Data*
A catalogue record for this book is available from the British Library

ISBN: 978-1-138-08283-0 (Set)
ISBN: 978-1-315-14775-8 (Set) (ebk)
ISBN: 978-1-138-08369-1 (Volume 13) (hbk)
ISBN: 978-1-138-08371-4 (Volume 13) (pbk)
ISBN: 978-1-315-11209-1 (Volume 13) (ebk)

**Publisher's Note**
The publisher has gone to great lengths to ensure the quality of this reprint but
points out that some imperfections in the original copies may be apparent.

**Disclaimer**
The publisher has made every effort to trace copyright holders and would welcome
correspondence from those they have been unable to trace.

# Mining and Development

FOREIGN—FINANCED MINES IN AUSTRALIA,
IRELAND, PAPUA NEW GUINEA AND ZAMBIA

CIARAN O'FAIRCHEALLAIGH

©1984 Ciaran O'Faircheallaigh
Croom Helm Ltd, Provident House, Burrell Row,
Beckenham, Kent BR3 1AT
Croom Helm Australia Pty Ltd, First Floor, 139 King Street,
Sydney, NSW 2001, Australia

British Library Cataloguing in Publication Data

O'Faircheallaigh, Ciaran
    Mining and development.
    1. Mineral industries    2. Investments, Foreign
    338.7'622         HD9506

    ISBN 0-7099-1937-9

St. Martin's Press, Inc., 175 Fifth Avenue, New York, NY 10010
First published in the United States of America in 1984

Library of Congress Cataloging in Publication Data

O'Faircheallaigh, Ciaran.
    Mining and development.

    Bibliography: p.291
    Includes index.
    1. Mineral industries – Developing countries – Finance
– Case studies.  2. Investments, Foreign – Developing
countries – Case studies.  I. Title.
HD9506.D452035  1984    338.2'3'091724     84-40049
ISBN 0-312-53367-5

Printed and bound in Great Britain

CONTENTS

To my parents, Tom and Una

LIST OF TABLES

# ABBREVIATIONS

| | |
|---|---|
| AAC | Anglo American Corporation Group |
| ABS | Australian Bureau of Statistics |
| AIDC | Australian Industry Development Corporation |
| AMAX | American Metal Climax, Incorporated |
| AMIC | Australian Mining Industry Council |
| APT | Additional Profits Tax |
| BCL | Bougainville Copper Limited |
| BCPL | Bougainville Copper Proprietary Limited |
| BMR | Bureau of Mineral Resources, Geology and Geophysics |
| CPD | Commonwealth Parliamentary Debates |
| CRA | Conzinc Riotinto of Australia Limited |
| IAC | Industries Assistance Commission |
| ICSID | International Centre for the Settlement of Investment Disputes |
| IBM | Irish Base Metals Limited |
| INDECO | Industrial Development Corporation Limited |
| LDC | Less Developed Country |
| LME | London Metal Exchange |
| MIMSU | Mining Industry Manpower Services Unit |
| MINDECO | Mining Development Corporation Limited |
| NCCM | Nchanga Consolidated Copper Mines Limited |
| NIF | National Investment Fund |
| PMA | Petroleum and Minerals Authority |
| PNG | Papua New Guinea |
| RCM | Roan Consolidated Mines Limited |
| RRT | Resource Rent Tax |
| RSG | Resources Study Group |
| RST | Roan Selection Trust Group |
| RTZ | The Rio Tinto-Zinc Corporation Limited |
| SCI | Smelter Corporation of Ireland |
| TPNG | Territory of Papua and New Guinea |
| UDI | Unilateral Declaration of Independence |
| UN | United Nations |
| UNIP | United National Independence Party |
| UPNG | University of Papua New Guinea |
| US | United States |
| WMC | Western Mining Corporation |
| ZIMCO | Zambia Industrial and Mining Corporation |

PREFACE AND ACKNOWLEDGEMENTS

This book is the result of a long-standing interest
in the economic and political issues raised by
foreign investment in mineral development. Its
contents reflect a conviction that these issues can
best be analysed through a comparative approach
which does not attempt too rigid a separation
between the economic and political aspects of
foreign mining investment. The research on which it
is based was conducted over the period 1976-81, and
was financed by the Australian National University
and the University of Papua New Guinea.
    Of the many mining companies I approached
seeking information and opinions only two, both of
which operate in the Republic of Ireland, refused to
have any involvement in the study. They did so
despite repeated requests on my part, and so any
error of fact or judgement regarding their
operations and due to their non-involvement is
entirely their responsibility.
    My thanks are due to the many mining company
and government officials whose assistance was
indispensable in undertaking this study, and to Ben
Smith, J.D.B. Miller, Ross Garnaut and Jeoff Jukes
who read and commented on all or part of an earlier
draft. I am also grateful to Christopher Helm for
his help in surmounting certain obstacles which
stood in the way of publication. Special thanks to
Oini Ume who typed the manuscript and to John Renaud
who helped with word processing.
    Above all my gratitude goes to my wife Carol,
who typed, proof read, and helped in many other
practical ways, and whose support and encouragement
were invaluable. And to my children, Sinead and
Kevin, whose frequent interruptions delayed the
appearance of this book but made its writing a good
deal more pleasant.


Ciaran O'Faircheallaigh

INTRODUCTION

This book attempts to identify, as accurately as possible, what occurs in and between countries when foreign investments are made in mineral development. In this regard, two areas are identified as requiring attention. (1) The nature of the transactions which constitute the process of foreign investment on the physical level. What is involved here are movements across national boundaries of objects, money and instruments of credit, information (interpreted in its widest sense to include technologies, skills and ideas), and individuals. (2) The nature of the relationships which are created between foreign investors and governments (or other relevant political identities) in countries where foreign investments are made. These areas can be dealt with separately for purposes of analysis, but they are closely interrelated in practice. The nature of physical transactions (or perceptions of the same) play a crucial role in determining the character of host country-foreign investor relations, and the policies and attitudes adopted by host country authorities exercise an important influence, in turn, on the physical effects of foreign investments.

There is of course little agreement as to what are the effects, beneficial or otherwise, of foreign investment. This lack of agreement is not entirely a result of differences in political outlook (though such differences are sometimes important). It results to some extent from a scarcity of 'hard' factual information. The literature abounds with checklists of perceived 'costs' and 'benefits',[1] but studies which present empirical data concerning particular cases of foreign investment are less plentiful. Where possible, such studies have been attempted in this book.

1

## Introduction

Uncertainty as regards the effects of foreign investments also reflects the existence of two methodological problems associated with study of the subject. The first involves the difficulty of attaching significance to particular impacts in individual countries. It is unlikely that a particular effect will have the same significance in countries with different economic, social and political characteristics. What criteria should be applied in assessing the significance of a certain impact in any given situation? The second problem arises from the difficulty of distinguishing between the impact of foreign-financed mineral development and mineral development per se. Such a distinction is required if a causal relationship is to be established between events in the host country and the specifically foreign nature of the mining investment which has occurred. Neither of these problems is susceptible to easy solution, but it is important to state explicitly what position is adopted in relation to both before embarking on a study of this kind, as the position taken can have an important bearing on the manner in which empirical data is interpreted. This is done in the first part of Chapter One.

That Chapter also outlines the kinds of effects which are considered in analysing the impact of foreign mining projects. It seeks to provide a broader perspective against which the analysis of individual projects can be conducted, by discussing the economic and other issues raised by foreign mining investment. One such issue involves the possible effects of foreign investors on political processes in the host country. As mentioned already, a close interrelationship exists between the policies pursued by host country authorities and the outcome of foreign investments, and this provides the background to the development of host country-foreign investor relations. Chapter One reviews some existing literature regarding the nature of those relations, and suggests some alternative perspectives.

A case study approach was adopted in order to permit detailed empirical studies of specific projects and of policies pursued by individual host country governments. The scope of the study was set to include countries characterised by varying economic, social, political and historical circumstances. The intention is both to examine the manner in which these circumstances affect the foreign investment process, and to discover what the

2

interaction between the two reveals about the nature
of that process. The countries examined are
Australia, Ireland, Papua New Guinea and Zambia.

Each case study outlines the manner in which
foreign mining investment has occurred, discusses
the mineral policies of the government concerned,
analyses the economic and other effects of foreign
mining projects, and attempts to assess the
appropriateness of host government policies and the
nature of host government-foreign investor
relationships. In the studies of Ireland, Papua New
Guinea and Zambia, a narrative approach is adopted;
developments relating to individual projects and/or
companies are outlined, and the analysis conducted
at appropriate stages of the narrative. It is
possible to adopt this approach because each of
these countries is host to only two or three major
companies mining one or two commodities, and
consequently detailed studies of individual projects
can be combined with an analysis of the broad
national and -international issues arising from the
position of the mining industry as a whole within
each country. This possibility is excluded in
Australia because of the wide range of minerals
produced and the large number of corporations
involved. An examination of a few companies or
projects in one or two sectors would be unlikely to
provide a basis on which to deal with national
issues, while constraints of time and space made a
comprehensive approach impossible. Thus in the
Australian case, government policies and the
assumptions on which they were based will first be
outlined, and the validity of those assumptions and
the appropriateness of policies subsequently
examined in the light of an analysis of the economic
and other effects of foreign mining investment as a
whole.

Before outlining its subject matter in more
detail, it may be useful to clarify what this book
does not attempt to do. First, for reasons explained
in Chapter One, it does not attempt an exact
calculation of every impact of foreign mining
investment with the aim of arriving at a single
figure which would stand for all economic and other
costs and benefits. Rather the aim is to identify
the various effects of foreign-financed mineral
development, assess their significance from the host
country's point of view, and isolate the factors
which determine the scale and direction of those
effects. Such an approach can yield useful
information regarding the character of foreign

# Introduction

mining investment, the kinds of policies which are likely to maximise its benefits and minimise its costs to the host country, and the nature of the foreign investor-host country relationship.

Second, my main concern is with what occurs when a foreign mining investment is made, not with what motivated that investment. Corporate motivation may of course have an important bearing on the outcome of an investment, and the issue is not ignored, but neither is it a focus of attention. However, one general point should be made regarding the motives of the companies encountered in this study. With few exceptions, they were not concerned to secure captive sources of raw materials for consumption in their home countries, but rather to apply their capital profitably in developing minerals for sale in markets outside their home countries. As will become apparent, this fact had important implications for the host countries.

Third, the book is not explicitly concerned with the nature of international trade in minerals, or with what this implies for the relationship between mineral exporting and mineral consuming countries. This issue is enormously important for both the host country and the foreign investor, as in most cases the value accorded minerals in international trade determines the size of the surplus available for distribution between the two. However, the focus of the book is on the manner in which that distribution occurs, rather than on the factors which determine its perimeters.

Notes

1. See, for example, Industrial Policy Group, The Case For Direct Investment (London, January 1970); S.J. Rosen and W..S. Jones, The Logic of International Relations (Cambridge, Mass., 1974), p 104; Z. Mikdashi, The International Politics of Natural Resources (Cornell University Press, London, 1976), pp. 42-3.

Chapter One

THE FRAMEWORK OF ANALYSIS

Some Methodological Problems

Two principal methodological problems were
encountered. The first, especially relevant in a
comparative study, arises from the difficulty of
assessing the significance of particular impacts in
individual countries. For example, provision of
employment in areas of Ireland suffering from
economic and social decline may obviously have
implications quite different to creation of
employment on a similar scale in remote and
sparsely-populated areas of Australia. In other
words, describing the physical impacts of foreign
investment does not automatically allow weight and
significance to be attached to those impacts, that
is to assign them as 'costs' or 'benefits' of
various magnitudes.
     Two alternative approaches could be adopted in
dealing with this problem. The first would involve a
calculation of the extent to which various impacts
have aided or obstructed host country governments in
pursuit of their policy aims. For instance,
generation of foreign exchange by a foreign mining
operation would be assigned as a 'benefit' where the
government was concerned to maximise foreign
currency inflow, with the significance of the
benefit being determined by the size of its
contribution in relation to total perceived needs.
Likewise if utilisation of foreign investment
diminished the possibility that a government's
desire to have minerals processed locally would be
satisfied, this impact could be assigned as a
'cost', whose magnitude would depend on the scale of
economic opportunities believed to have been lost.
     Adoption of this approach would create a major
difficulty. It would involve acceptance of

government policies as given, excluding any critical element from the analysis. This would be undesirable for at least two reasons. First, it would diminish the practical value of the study, which would merely consist of a catalogue of policies, a catalogue of impacts, and a comparison between the two. No conclusions could be reached regarding the appropriateness of the policies being pursued. In other words, no recognition would be given to the fact that policies can play a role in <u>determining</u> the impact of foreign mining investment, and that policy-makers should take account of the likely or actual effects of such investment in formulating policies. For instance, it might be stated that only a small percentage of the foreign exchange earned by a foreign mining project remained in the host country, and that the project consequently contributed little to government endeavours to overcome balance of payments deficits. But the important issues in this case (both in terms of their inherent interest and in terms of their implications for policy-making) are, for example, whether a <u>higher</u> proportion of foreign exchange could have been retained if different policies had been adopted, or whether a low level of net foreign exchange inflow is inevitably associated with this type of project? If the latter, could the host country pursue a more sensible strategy in attempting to balance its external transactions?

Second, a crucial issue in examining host country-foreign investor relations involves the possibility that foreign investors might influence host country authorities into adopting policies which are in fact not likely to maximise host country returns from mineral development. From this perspective also it is essential to assess the appropriateness of government policies.

A second approach would be to inquire whether foreign mining investment had contributed to a more efficient utilisation of resources (by mobilising idle resources or by attracting resources from alternative, less productive uses), or whether it had the opposite effect. This approach could lead to quite different conclusions from the first. For example, it might transpire that mineral processing would represent a relatively inefficient use of resources, and the failure of foreign companies to invest in processing would consequently not be regarded unfavourably. This second approach is essentially that of the economist, who attempts to establish optimum patterns of resource allocation, and assesses particular economic activities by gauging the extent to which they contribute to or detract from achievement of the optimum. However, this study is concerned with both economic and political aspects of foreign mining investment, and such an approach would not readily facilitate the political element of the analysis. Thus a compromise approach has been adopted. Attention is focused on the extent to which foreign mining investment accords with host government policies, but in addition the appropriateness of those policies is assessed in the light of economic and other circumstances prevailing in individual host countries, of their effect on foreign mining investment, and of the actual impact of such investment. This assessment frequently does involve judgements as to the efficiency of allocating resources in one way rather than another, but does not require a rigorous economic analysis aimed at establishing optimum patterns of resource allocation.

The second methodological problem arises from the need to establish a causal relationship between the occurrence of foreign mining investment and empirically-observed phenomena in the host country. In one sense, this relationship can be established quite easily, as foreign mining projects are clearly responsible for particular physical occurrences. However, the question arises as to whether the foreign nature of the investment concerned is responsible, or whether those occurrences would result from any type of mineral development? To answer this question, we require a basis of comparison which, by illustrating what would happen in alternative situations, would permit the effect of the specifically foreign nature of an investment to be isolated. However, serious difficulties arise in devising an acceptable basis of comparison.

One approach would be to ask what was the most likely alternative to foreign investment in each individual case, to indicate the likely outcome had this alternative materialised, and to compare that outcome with the actual impact of foreign investment. Three major problems would be associated with such an approach. First, it is frequently unclear what was the most likely alternative to foreign investment in particular instances. This applies, for example, in the case of Australia. A number of Australian individuals and corporations have involved themselves very successfully in mineral development, and it could be argued that, in the absence of foreign investment, additional domestic resources would have been mobilised and all viable mineral deposits exploited.[1] The most likely alternative to foreign investment would therefore be domestic investment. On the other hand, it could be the case that existing Australian companies have absorbed all, or nearly all, of the relevant resources (e.g. risk capital, technical and managerial skills) available domestically, and that the most likely alternative to foreign investment is no investment. The truth probably lies somewhere in between, but precisely where?

Second, even if it could be stated with certainty that domestic investment *was* the alternative to foreign investment, what assumptions are made about the character and behaviour of the domestic investors? Would their mining operations have been as efficient as those of the foreigners who did invest? Would they have enjoyed similar market outlets, or applied similar criteria in assessing prospective investments? Third, at any particular time the alternative to foreign investment frequently *is* total lack of investment,[2] and consequently there is no basis of comparison on which to identify the effects of the foreign nature of the investment which did occur.

A second approach would be to posit some hypothetical form of investment which would then be used as a basis for comparison, an approach frequently adopted, though usually implicitly, in the literature on foreign mining investment in developing countries. For example, a comparison is made between the impact of the foreign investment and the expected outcome had mineral development been undertaken by a highly-efficient public mining corporation acting to maximise benefits and minimise costs from a national perspective; divergences between the two are thought to represent the effect of the former's foreign nature.[3]

This approach also creates difficulties, in that it requires assumptions regarding the properties which would be attributed to various kinds of hypothetical investment. In the literature, such assumptions usually reflect the political inclinations of the individual involved rather than any basis in fact. It is assumed, for example, that private enterprise would be inherently more efficient than public enterprise, or vice versa. My research produced little empirical evidence to support contentions of this type; as regards this specific example, efficient and inefficient private mining enterprises and efficient and inefficient public enterprises were encountered. In addition, the comparison frequently does little to illustrate the impact of the foreign nature of an investment, because of the type of assumption made, a point illustrated by the example quoted in the previous paragraph. No private investor could be expected to undertake activities which, though maximising benefits from a national perspective, might result in financial loss to himself. Thus much of the comparison is in reality between the hypothetical investment and any form of private investment.

In fact it proved impossible to devise a general basis of comparison which would illustrate the effects of the foreign nature of foreign mining investments. All that can be done is to make clear in the case studies precisely what type of causal relationship is being claimed and, where comparisons are made, to make explicit the assumptions on which they rest.

## The Effects of Foreign Mining Investment

As mentioned in the Introduction, this book does not attempt to arrive at a single figure which would express the sum of the economic and other costs and benefits of foreign mining investment. While it is important to gauge the various effects of foreign investment as accurately as possible, I do not believe it is feasible to arrive at such a figure. Certain costs or benefits cannot be quantified, the significance of a particular impact can vary from case to case, and assessment of certain effects requires value judgements which only policy-makers in the host country are qualified to make. And though these difficulties might be mitigated if the analysis was limited to purely economic effects, the methods available for calculating a figure or set of figures to express those effects have important drawbacks from the point of view of a study such as this.[4] First, they involve simplifications and assumptions[5] which, while unavoidable in formulating and applying economic models, would here be unacceptable and indeed might assume away issues crucial to this study. So, for instance, in outlining a cost-benefit scheme for appraising foreign investment, Bos et al assume 'that the government of the host country is able to pursue rational economic policies'.[6] But a major concern of this study is to analyse the factors which determine whether or not a government will pursue rational economic policies and to examine the impact on

9

foreign mining investment of its ability or
inability to do so. Second, the calculation of a
single figure which expresses the economic impact of
foreign investment often reveals little about the
economic and political processes which led to this
particular outcome.[7] So, for example, it is possible
to develop a model which will permit calculation of
the balance of payments effects of a foreign investment;[8] the model might indicate a negative impact on
the balance of payments, but would not show why such
an impact materialised. It might be, for example,
that government taxation policy resulted in a very
high proportion of profits being remitted by the
foreign investors, leading to large outflows of
foreign exchange; the model would give no indication
of a connection between that policy and the outcome
indicated. These comments are not meant to imply a
general rejection of model-based, quantitative
methods of analysing foreign investment, but merely
to note their inadequacy in a situation where the
principal concern is the interaction between public
policy toward foreign mining investment and the
economic and other effects of such investment.

The approach adopted is to identify as
accurately as possible, and to discuss, the effects
of foreign mining investment under various headings
which correspond to issues raised in the literature
on such investment and on foreign investment
generally. It is of course impossible to assess each
and every effect of foreign investment, and the
discussion is confined to issues which figure
prominently in both the literature and in the
concerns of host country authorities. Five general
kinds of effects that have arisen, and have been
considered separately, in the literature are dealt
with.[9] First, such investments may generate incomes
for nationals, by creating employment, providing a
market for goods and services, providing raw
materials for mineral processing, or generating
economic rents which can be taxed by government.
These incomes may accrue to the host country as
foreign exchange, allowing it to finance imports of
goods and services. Second, the investment process
may involve provision of resources which are in
scarce supply in the host country, particularly
capital, technology, and technical and other skills.
Third, foreign mining projects may create external
economies which allow nationals to enhance their
incomes, for example if 'spill-over' of technology
or technical expertise allows domestic enterprises
to operate more efficiently, or if mine

infrastructure is utilised in other economic
activities. Fourth, mining activity may create
external diseconomies, by absorbing resources which
have alternative uses, by using inappropriate
technology, or by adversely affecting consumption
patterns or domestic income distribution. Finally,
foreign mining investment may have politically
undesirable effects. Host country sovereignty may be
undermined, or foreign investors may subvert
domestic political processes to ensure policy
outcomes favourable to themselves.

## Generation of National Incomes

Perhaps the most apparent impact of foreign investment is that the economic activities it allows
create financial surpluses. One of the principal
areas of interest in the literature is the question
of how and to whom these surpluses accrue. Do they
accrue entirely to the foreign investor? If not, how
are they distributed? How should they be
distributed? Can foreign investors remove surpluses
from host country jurisdiction and so skew the
distribution in their own favour?

      These questions are particularly relevant in
the area of mineral development, as economic rents
or 'profits in excess of an appropriately risk-
adjusted rate of return on funds employed'[10] are
frequently substantial. The ability of mineral
producers to exact such rents results from the
scarcity, current or future, of mineral resources
(reflecting their non-renewability and their uneven
distribution), from the imperfect nature of
competition in mineral industries, and from the fact
that mineral deposits vary considerably in richness,
accessibility, and ease of exploitation.[11] This last
factor means that while marginal deposits will, by
definition, merely generate sufficient surplus to
ensure continued investment and production, non-
marginal mines can generate considerably greater
surpluses. In sum, mineral exploitation can generate
profits far in excess of those required to ensure
continued development.

      This situation provides host country govern-
ments with an opportunity to enhance revenue
substantially and there are strong arguments for
asserting that economic rents should accrue largely,
if not entirely, to the host country.[12] One such
argument is ethical: since mineral resources are
generally believed to belong to society as a whole,
it is felt that profits in excess of necessary

payments to factors of production (e.g. wages,
interest payments, operating expenses, appropriate
returns to capital) should accrue to the community.
In economic terms, the concept of tax neutrality
(discussed in footnote[13]) offers a strong incentive
for appropriation of economic rents.

Thus it is important that host country
authorities apply appropriate taxation arrangements
to foreign mining projects. Two general
considerations are relevant in formulating such
arrangements – to maximise government revenue, and
to avoid detering investment or distorting
production decisions. These requirements will
frequently conflict, and the extent to which they
can be reconciled depends importantly on the _manner_
in which taxes are collected.

Profit-based taxes which reduce the post tax
rate of return below the rate required by foreign
investors, or are expected to do so, will deter
investment; taxes which are a cost on production
(unit or ad valorem royalties)[14] may also distort
production decisions. In theory, these effects are
avoided where revenue is raised solely by taxation
of economic rents. By definition, such taxation will
not reduce the post tax rate of return below that
required by foreign investors, and neither will it
affect production decisions since it will have no
impact on the point at which marginal costs equal
price. However in practice application of rent
taxation by individual governments may not be
neutral in its effects on investment and production
decisions. Competition among host country governments for foreign investment may lead one or more of
them to forego rents and, where the access of the
foreign investors to capital and/or to market
outlets is limited, application of rent taxation by
other countries may deter investment and/or lead to
a decline in utilisation of existing capacity.

In addition, practical difficulties arise in
devising a tax system which will capture a high
proportion of economic rents and will capture only
economic rents. Governments operate with imperfect
knowledge, and rarely know what _is_ 'an appropriately
risk-adjusted rate of return'. The rate may be set
too high, allowing rents to escape taxation and
encouraging wasteful investment, or too low,
bringing the post tax rate of return below that
required by the investor. There are also important
political constraints which militate against
adoption of a purely rent-based tax system.
Governments are rarely certain in advance that rents

will materialise, and neither they nor their
constituents are prepared to accept the possibility
that their country may not obtain any financial
returns from exploitation of its mineral resources.
Consequently, profit or production-based taxes are
usually applied. This of course increases the
foreign investor's uncertainty as to the outcome of
his investment, in other words it increases the
variance of possible outcomes, leading him to demand
a higher expected rate of return.[13] As a result,
certain projects may be rendered marginal, and
economic rents which might otherwise have
materialised may be reduced or eliminated.
Governments may be willing to accept this situation
because of a preference for an expected income
stream which, though lower, is more certain.

Finally, it should be noted that the above
discussion relates to situations in which
governments are seeking new investment, whether by
operating companies or by companies which will base
their investment decisions on the expectation of
being treated in a manner similar to such companies.
A substantially lower rate of return will be
required to ensure that existing mines continue to
be operated efficiently, and where no new investments are expected, either by way of expansion of
existing mines or establishment of new mines, a
government may be able to capture a very high
proportion of total profits.

Thus the important issue is the extent to which
host country taxation systems reconcile the need for
revenue with that for further investment and
efficient mineral exploitation, given the
circumstances which prevail in a particular country
and given the various constraints under which taxing
authorities operate. Each case study outlines in
detail the taxation policies of the host government
involved and the political and other considerations
which have helped shape them, examines their effects
on foreign mining investment, and attempts to assess
their appropriateness in the light of circumstances
in that country and of the earlier general
discussion.

Another issue which arises in relation to
mineral revenues is the possibility that foreign
mining companies might avoid taxation by disguising
part or all of their profits and removing these sums
from host country jurisdiction; in other words they
may be in a position to _define_ 'taxable profit'. The
most commonly-discussed method of doing this is by
transfer pricing. Put simply, this involves the

ability of corporations which control more than one stage of the productive process (i.e. are vertically integrated) to set their own prices for transactions between affiliates. Where a foreign investor can engage in this practice, he has the ability to define profitability at each stage of the productive process and, where these stages occur in a number of countries, can determine the point at which profits will appear as taxable surpluses. If conditions in a host country warrant it (e.g. high levels of taxation, restrictions on profit remittance), he can remove profits to another location. Transfer pricing creates particular difficulties for host countries as the lack of an 'open' market often denies them any reference point by which to judge the 'real' value of the commodity concerned.[16]

Another way in which incomes may accrue to nationals is through creation of employment, seen by many policy-makers as one of the most important benefits of foreign mining investment. Opponents of such investment argue that foreign-financed mining projects tend to be capital rather than labour intensive, and so employ relatively few people.

Two distinct questions arise in relation to the capital intensity of mining operations. First, could resources utilised in mining have been employed in alternative, more labour-intensive uses? If not, then from this perspective any employment resulting from foreign mining investment represents a net gain in relation to a situation in which that investment did not occur. Second, could or should mining operations themselves be more labour-intensive? It is important to clarify what is involved here. In theory, it is possible that use of alternative mining methods or technologies could allow substitution of labour for capital without affecting the size of economic rents. In practice, it is very unlikely that this would be the case - in any given situation, substitution of one factor for another would probably increase or diminish economic rents, depending on relative factor costs. Thus a host country must decide whether its aim is to maximise economic rents, in which case it will allow foreign investors to employ profit-maximising factor combinations, or whether it wishes to maximise employment, in which case it must accept that, in certain circumstances, economic rents will be diminished.

Foreign mining investment may also add to national incomes by creating additional economic activity through linkage development. Proponents of foreign investment frequently view projects it finances as 'engines of growth' which induce economic expansion elsewhere in the economy by creating a demand for goods and services (backward linkage) and by supplying inputs to other industries (forward linkage).[17] Alternatively, foreign-financed projects are seen as 'enclaves' which are located in the host country in a geographical sense only, and which, because of their pattern of sales and purchases, are integrated into the international corporate economy of affiliated or associated companies rather than into the domestic economy.[18] The case studies outline the extent to which linkage development has actually occurred, and attempt to assess the degree to which the foreign nature of mining investment has inhibited linkage growth, if at all. The appropriateness of host country policies towards linkage development is also considered.

For many host countries, a positive contribution to the balance of payments is regarded as the principal justification for encouraging foreign investment. The foreign exchange earnings represented by such a contribution are believed to be indispensible if imports of essential goods and services are to be maintained. A favourable balance of payments effect can emerge from an excess of capital inflow over outflow of capital and interest, or from the earnings of foreign-financed export industries. Critics point out that the long-term effect of foreign investment may in fact be a balance of payments deficit and, where this is not so, that the net balance of payments effect may be very much smaller than gross foreign exchange earnings.[19]

A number of authors have questioned the validity of attaching importance to the balance of payments impact of mining projects as such. They point out that this effect is largely determined by the extent to which incomes generated by mining accrue to domestic factors of production and to the taxing authority. As B.L. Johns has commented in relation to Australia,

> [if] we have already taken account of the various ways in which the incomes of Australians might be increased as a result of overseas exploitation of our mineral wealth, it would be inappropriate to allow for the balance of payments effect as an additional factor.[20]

There has been a fundamental failure in some host

countries to comprehend fully the link between the impact of foreign investment on domestic incomes and its effect on the balance of payments, principally because creation of export income has been identified as the relevant criterion in examining this latter effect. This failure has resulted in the anomalous situation in which the granting of generous taxation concessions to foreign mining companies has been justified in terms of the export receipts which mineral development is expected to generate, whereas such measures themselves reduce the amount of foreign exchange available to the host country for purchases of essential goods and services.

In each case study an attempt is made to calculate the net foreign exchange inflow resulting from foreign mining operations. This is done by calculating the difference between the inflows of capital and export receipts and outflows of dividends, interest payments, debt repayments, undistributed profits, and direct and induced imports; that difference equals the total of domestic incomes, or the 'retained value' of mining operations. An attempt is also made to identify the various components of national income as accurately as possible. This exercise illustrates the link between the impact of foreign investment on national incomes and on the balance of payments, while the identification of individual categories of foreign exchange outflows and of the composition of retained value will provide useful evidence on which to assess the impact of foreign mining investment and the appropriateness of government policies.

Provision of Scarce Resources through the Investment Process

The most obviously apparent advantage of foreign investment in this regard is that it provides capital in situations of inadequate supply, whether these result from an absolute shortage of domestic capital or from inefficiencies in its allocation. Proponents of foreign investment argue that host countries can also benefit substantially from transfers of technology and technical and other skills, especially where processes or skills are passed on to domestic companies and/or to other sectors of the economy.[21] Those opposed to foreign investment argue that the type of technology introduced may be inappropriate (e.g. by emphasising use of scarce capital as opposed to plentiful labour),[22]

16

and that the skills imparted rarely have application outside the foreign investor's own industrial process.[23]

The accuracy of these opposing viewpoints will be examined in the case studies, but at this stage one general point should be made. The application of technologies and skills by foreign investors can only be regarded as a benefit from the host country's point of view to the extent that it captures the fruits of the more efficient utilisation of resources which results. Where the effect is to render otherwise uneconomic mineral deposits marginal, the benefit to the host country depends on the extent to which local factors of production are employed. Where economic rents are generated, it also depends on the degree to which these are captured by the taxing authority.

## Creation of External Economies

In addition to any 'spill-over' of technology or expertise, foreign mining investment may create external economies by providing infrastructure which can be used in non-mining activity.[24] Activities which would otherwise be uneconomic may be rendered viable, or incomes from existing activities may be increased because freight, power or other costs are reduced. In some cases governments construct infrastructure and may or may not charge mining companies for its use. Where charges are made the question arises as to whether these fall short of, or exceed, the cost of providing infrastructure. If the former is the case or if infrastructure is provided free of charge, the rationale for such a policy requires assessment.

## Creation of External Diseconomies

A number of external diseconomies may be associated with foreign mining investment. Foreign investors may absorb resources which could have been utilised in other economic activities, creating opportunity costs which must be allowed for in assessing their contribution to national incomes. For instance, foreign concerns sometimes utilise precisely those sources of local capital whose paucity supposedly justifies their presence. Two major studies of foreign investment in general[25] have concluded that this practice is widespread, and a third study has estimated that US companies operating abroad raise 60 per cent of their capital locally.[26]

17

Also relevant in this context is the fact that foreign mining investment absorbs non-renewable mineral resources. In this instance an opportunity cost may also be created, in that domestic investors who might later have developed the capacity to exploit those resources are denied the opportunity to do so. The point at issue here is whether the immediate benefits of foreign-financed mineral development outweigh the future benefits of domestically-financed mineral development, after application of an appropriate discount rate to the latter. This is an extremely complex issue, as it demands assumptions regarding movements in prices and production costs for individual minerals over time, assumptions which can seldom be made with very much certainty or accuracy. Other issues are also involved. What changes are expected to occur over time in the magnitude or significance of any spin-off effects anticipated from mineral development? To precisely what extent and over what time period is domestic expertise expected to develop? Political factors are also relevant. For example, a government attempting to establish its legitimacy in a newly-independent country or simply trying to retain office may place a high value on benefits expected to accrue in the short term.

It would clearly be impossible to reach precise conclusions regarding the economic advisability of 'leaving minerals in the ground' in any given situation. Thus in the case studies this issue is dealt with in general terms only, and it is in that context that the appropriateness of the time horizons actually employed by decision-makers will be assessed.

Foreign investors may also create external diseconomies by encouraging inappropriate consumption patterns in the host country. Consumption patterns may be deemed inappropriate on a number of grounds, for instance if they result in unnecessary use of resources which could profitably be employed elsewhere (e.g. use of private cars rather than public transportation), or if they lead to expenditure of valuable foreign exchange on imported consumer goods. Distortion of consumption patterns is likely to be associated with consumer goods industries which devote a high proportion of their expenditure to advertisement. However, mining companies can have an impact if their employees represent a visible and well-defined group which exerts a 'demonstration effect' by practicing consumption patterns of a certain kind.

Finally, external diseconomies may result from the impact of foreign mining investment on domestic income distribution. The creation or worsening of disparities in wealth between different sections of the community can be criticised on various grounds. One is moral, and involves the claim that such disparities are inherently evil. Others are political and economic in nature. For example, if small groups of wage earners become wealthy in comparison to the wage-earning population as a whole, one of two equally undesirable consequences may result. Should general wage levels remain unchanged, the income differential between the privileged group and other wage earners may give rise to economic and political disruption. If the 'demonstration effect' exercised by the wealthy few pushes general wage levels upwards, economic development may be hindered (for example because economically marginal activities are rendered uneconomic). Regional inequalities in income distribution can also create serious political tensions, particularly in newly-independent countries which lack a strong sense of national identity.

Foreign mining investment may affect domestic income distribution in a number of ways. First, it may reduce (increase) returns to domestic capital by pre-empting (facilitating) its involvement in mining operations which yield a higher return than alternative investment opportunities. Second, it may increase (decrease) returns to labour by influencing wage levels, either directly or indirectly. Third, it may increase (decrease) disparities in regional income distribution, depending on the geographical pattern of its expenditure.

A substantial number of variables are involved in determining the effect of foreign mining investment on wage levels and on income distribution generally. The manner in which they interact in individual host countries is examined in the case studies, and the impact of foreign mining projects assessed accordingly.

The Political Impact

Foreign investment is frequently thought to result in loss of host government sovereignty, a concept which covers a number of issues. Foreign investors are believed to be capable of taking decisions which affect the national welfare and, in doing so, of ignoring the wishes of host country governments.[27] The decisions involved might relate, for example, to the location of investments, retrenchment decisions, circumvention of credit restrictions, withdrawal of funds in times of balance of payments problems, or conclusion of takeover agreements.

A distinction must be made in this area between the potential power of foreign investors and their behaviour in practice. Foreign companies may well refrain from exercising any power they may have in order to maintain good relations with the host country over the longer term. In addition, it is apparent that the potential of various investing companies in this regard will vary with the nature of their own institutional frameworks.[28] Where subsidiaries enjoy considerable autonomy and are linked to the parent company primarily through use of a common technology or trade name, little difference may exist between their responsiveness to host government policies and that of domestic companies. Where subsidiaries supply components to a productive process spanning national boundaries, responsiveness may be very low. Between the two extremes, many variations will occur. Historically, subsidiaries in the extractive industries have been very closely controlled by parent companies, reflecting the need to ensure supplies of high-quality products on a reliable and punctual basis. The question of national control over corporate decision-making has in fact figured very prominently in the case study countries, and it is consequently dealt with at various stages of the narrative rather than as a separate issue.

Host government sovereignty is also seen to be threatened where home country governments intervene on behalf of foreign investors or where foreign investors are used as an arm of home government foreign policy. In this latter regard companies operating abroad might be passive (and even reluctant) tools of home government policy, being constrained from acting in certain ways, for example by anti-trust laws or by regulations prohibiting trade with particular countries. Alternatively, they might take positive initiatives on behalf of the home government. Finally, there is the possibility that foreign investors might subvert political processes in the host country. This issue brings us to an area which can more properly be dealt with in the context of relations between host country governments and foreign investors.

The Framework of Analysis

Foreign Investor-Host Country Relations

This discussion emphasises the fact that policies
adopted by host country authorities exert a crucial
influence over the impacts which will result from
foreign investment. The clearest example of such
influence is in the area of economic rent
distribution, where control over taxation assigns a
central role to government. Taxation policies, in
turn, influence the distribution of benefits
arising, for example, from transfers of technology
or technical skills, and help determine the balance
of payments impact of foreign mining projects. In
addition, host country policies in relation to
linkage development and provision of infrastructure
can be of considerable significance. On the other
hand, the effects of mining investment, real or
perceived, exert a major impact on government
policies, as will become apparent in the case
studies. This situation provides the background for
the development of relations between foreign
investors and host governments (or other entities
with influence over the decision-making process),
and it is to the analysis of these relations that I
now turn.
   Much of the existing analytical work dealing
with host country-foreign investor relations in
mineral industries concentrates on explaining
changes in the relative bargaining positions of host
countries and foreign investors (frequently within a
framework of game theory).[29] Raymond Mikesell, for
example, has outlined the way in which bargaining
power shifts from the foreign investor to the host
government as development of a deposit proceeds. In
general, the bargaining power of the foreign
investor is greatest before a new investment is made
and lowest after such an investment is completed and
has proved profitable.[30] Host governments, initially
faced with the overwhelming bargaining power of the
foreign investor, accept mining agreements which are
very favourable to the latter. Once the investment
has been made and becomes profitable, however, a
shift in bargaining positions takes place. The
mining venture has made its commitment, it cannot
remove its physical assets (as can, for example,
certain kinds of manufacturing concerns), and is
therefore unlikely to withdraw while it is allowed
even a 'modest' return on its capital. Political
pressure mounts on government to utilise its
improved bargaining position to increase its share
of the (now very apparent) surplus generated by

mineral development. It will usually do so by
demanding renegotiation of the original agreements.
Raymond Vernon has described this phenomonen as the
'obsolescing bargain', and has analysed the manner
in which the characteristics of individual metal
mining industries and of particular host countries
affect the degree to which shifts in bargaining
power occur in specific cases.[31]

Mikesell argues that the main problem caused by
such developments is one of controlling conflict
between the two parties. Conflict is regarded as
inevitable, but it can be mitigated to a very large
extent, for example by including provisions for
renegotiation in the original agreement or by
allowing for higher state revenues in the event of
project profitability being unexpectedly high.[32]
Severe conflict is in fact destructive to both
parties, as it may lead to reduction in the overall
surplus available for distribution, an outcome which
would be in neither of their interests.[33] Development of such conflict (or what Mikesell terms
'warfare situations') therefore results from
decisions which are irrational on an economic basis,
and involve 'a political act usually taken in
response to an ideological or other non-economic
motivation'.[34] The main problem is to discover
'joint maximising rules' which will allow both
parties to maximise their returns from mineral
development without undermining their basic
relationship.[35]

T.H. Moran, in a study of foreign investment in
the Chilean copper industry and in other
publications,[36] argues that this approach is
inadequate. Moran accepts that important short-term
shifts in relative bargaining strengths do take
place in the manner described above.[37] But he claims
that a simple bargaining model which outlines such
changes and recommends strategies for conflict
avoidance and 'joint maximisation' does not allow a
satisfactory analysis of relations between foreign
investors and host countries. His criticism is based
on the contention that such a model ignores the
dynamic elements involved in the whole political,
economic and technical framework within which bargaining takes place.

Moran identifies three such elements as being
particularly important: (1) changes in perceptions
of risk and uncertainty over the long term, (2)
changes in the monitoring and negotiating skills of
host country governments and in their ability to
carry out functions performed by the foreign

investor, arising from progress along what is described as a 'learning curve', and (3) domestic
political considerations resulting from the
interplay of local political and interest groups and
from changes in domestic perceptions of the role
played by foreign investors.

Moran points out that the initial foreign
investors expend large amounts of capital with substantial risks of failure before they collect
sufficient information to reduce uncertainty
substantially (e.g. regarding the structure of production costs, the 'operating climate' in a
particular country). In addition, these investors
have a monopoly over the capacity to discover and
develop mineral deposits. The heavy premium they
charge for uncertainty and their exercise of
monopoly power lead foreign investors to demand very
favourable conditions before agreeing to commit
funds. Host governments are not qualified to gauge
the real extent of risk and uncertainty, and therefore must accept the assessment made by the investor
and concede his demands. But once initial investments succeed, uncertainty as to the existence of
orebodies and as to the technical and economic
aspects of their development is greatly reduced. The
host government can consequently drive a tougher
bargain with subsequent investors, and may well use
the leverage thus gained to renegotiate earlier,
more generous agreements. Thus changes in bargaining
positions which accompany individual project
developments are not self-contained, but rather are
associated with a cumulative effect which serves to
improve the overall bargaining position of the host
country.[38]

The second element involves the concept of a
'learning curve'.[39] A country will initially be
lacking in the technical, accounting and economic
skills needed to analyse the proposals and, later,
the operations of international mining enterprises.
In any case, its government will not possess the
bargaining skills required to put such expertise to
good use. However, the financial success of initial
investments provides an incentive to develop capabilities in these areas. Thereafter, the host government can monitor the activities of foreign concerns
more closely. In addition, a monopoly of relevant
skills is no longer enjoyed by the foreign investor.
Government can push home the advantages derived from
these changes, using the bargaining skills it has
developed over time to impose stricter conditions on
foreign investors.[40] In combination with the

reduction of uncertainty mentioned above, such
progress along the 'learning curve' represents 'a
fundamental evolution in the balance of power
between the international investors and the host
country', and it is within this 'slow but cumulative
shift in the balance of power' that rapid short-term
shifts in that balance take place.[41]

In this scheme of things, conflict is not a
problem to be avoided or mitigated, but rather
reflects the struggle by a host country to improve
its bargaining position and to utilise the advances
made in this area. Neither do attempts to gain
control of mining industries reflect 'irrational'
decisions taken for non-economic reasons. 'There is
in fact a rational pattern of technology transfer,
learning and interest articulation beneath the
apparent "waves" and "emotions" of economic
nationalism' and the actions of economic
nationalists represent 'a systematic drive to take
over functions, gain expertise, acquire knowledge
that were previously the monopoly of the foreign
corporations'.[42] The rational strategy for the
economic nationalist is to

> take as full advantage of the foreign corpor-
> ations as he can while they are within his
> range, and not to push them out until he is as
> close as possible to being able to duplicate or
> surpass their feats on the international as well
> as the national scene.

But pushed out 'they' will be: 'such foreign
corporations in the long run, to paraphrase Keynes,
are all dead'.[43]

Moran argues that by examining the degree of
uncertainty prevailing and the extent of progress
along the 'learning curve', one can gauge 'the array
of options open to policy-makers within the horizon
of domestic knowledge and experience at each stage
of relations with the foreign investors, and to
measure domestic performance against these options'.
In fact, there are many instances in which available
options are clearly not realised. These are
explained by pointing out that policy is not formulated by a monolith entitled 'host country', but
rather is 'the outcome of the struggle of diverse
groups, of successive administrations and their
adversaries, to maximise their own conceptions of
the national interest through manipulation of
[mineral] policy'.[44] Thus the third dynamic element
is introduced. Political interactions of this type

explain 'deviations' from the behaviour one would expect from a notional entity (the 'host country') acting to maximise the potential returns arising from its 'objective' bargaining position.

At certain stages, domestic political and interest groups with influence over the decision-making process will see it in their interest to make economic or political 'pay-offs' to foreign investors by not exercising the state's full bargaining power. Moran explains in detail the reasons for occurences of this type at various times in the history of the Chilean copper industry.[45] He defines 'exploitation' as the extent to which a host country suffers financial loss as a result of not exercising its full bargaining power. An 'asymmetry of power per se' between host countries and foreign investors is not viewed as a basis for exploitation, but rather only 'the refusal of domestic elites to act in the national interest to the extent that they have the power to do so'. There is no 'exploitation' without complicity on the part of domestic elites.[46]

The concept of elite complicity is of course found in much of the literature on foreign investment in developing countries, and it is through such complicity that foreign investors are usually thought to subvert host country political processes. The elite's willingness to sacrifice the national interest is variously attributed to outright bribery by foreign investors,[47] to its belief that foreign support is essential for maintenance of a state system within which it is the primary beneficiary,[48] or to the desire of a certain section of the elite to enlist foreign help in maintaining its political dominance. Moran favours this last explanation, but stresses that there is no permanent mutual interest between local political groups and foreign investors in 'exploiting' the host country. (Theories of 'neo-imperialism' which posit such an interest are regarded as static and incapable of explaining change in foreign investor/host country relations. Such theories pick one point in those relations and wrongly assign permanence to that situation.)[49] Rather elite complicity is viewed as a temporary departure from a long-term trend which is leading to greater domestic returns from, and control over, mineral industries. Pursuit of this trend is not the preserve of any one political or ideological grouping. Moran points out that the Chilean Conservative Party, which was at the centre of the major 'pay-offs' to foreign investors, eventually perceived its interests as different to those of US mining companies. It joined with other political parties of the centre and left in 'squeezing' those companies, and eventually in their nationalisation.[50]

Moran seems to assume that such an outcome was inevitable. Political demands for greater domestic returns from, and control over, the copper industry, provided Chilean leaders with the impetus for progress along the 'learning curve'. Such progress, combined with reduction of uncertainty, put Chile in a stronger bargaining position. It also reduced elite perceptions of the cost of dispensing with foreign investment and increased perceptions of the benefits to be gained from doing so. Elites of all ideological hues therefore inevitably bowed to the political imperatives, and took action to improve Chile's position. According to Moran, this type of progression was, or is, not confined to Chile or to the copper industry, but occurs in other countries and in the oil, ferrous and non-ferrous metals industries.[51]

Thus Moran both offers an explanation for empirically-observed patterns of behaviour and predicts that host country-foreign investor relations will follow an inevitably conflictual course, resulting in the eventual withdrawal of foreign investment. The case studies will provide empirical evidence on which to judge the validity of his propositions, but at this stage a number of general points can be made.

Three relate to the concept of progress along a 'learning curve'. First it should be remembered that 'a systematic drive to take over functions, gain expertise and acquire knowledge' may impose substantial costs on the host country, absorbing human and financial resources which have alternative uses. Moran assumes that the benefits of following such a course will outweigh the costs, but this appears to be a matter for investigation rather than a justifiable a priori assumption. Second, he does not consider the possibility that a host country may exaggerate its ability to 'duplicate or surpass' the feats of foreign investors. It is possible that host country authorities might overexercise, as well as underexercise, their bargaining power. They may miscalculate on their ability to maintain production efficiently after the foreign investor has been 'pushed out', and his departure may consequently impose considerable costs on the host country. Third, it is by no means self-evident that a host country will inevitably move along a learning curve even if it is in its interests to do so. There are a number of aspects to this last question, the first of which concerns Moran's contention that elite complicity will necessarily be temporary.

It is not obvious that this will necessarily be the case. First, elites may have a continuing interest in engaging in complicity; as Moran himself points out, some very specific factors combined to persuade Chile's conservative elites to abandon their alliance with US multinational mining companies, and it cannot be assumed that similar circumstances will apply in other countries. Second, Moran assumes operation of democratic structures in the host country. There is in fact no guarantee that domestic elites will respond to public pressures for termination of exploitative relationships.[52] Third, even where democratic structures do exist, pressure for termination of complicity may not materialise. Foreign investors may merge into the local landscape over time, and this may serve to reduce political pressure against elite cooperation. In addition, politicians may not have a monopoly over the ability to grant concessions. These may also result from bureaucratic processes which are secret in nature, and consequently may not attract public attention (for instance interpretation of tax legislation by the enforcing authority). Finally, it is possible that a mining industry, though significant in national economic terms, may not be prominent in public consciousness, and for this reason also complicity may not be detected.

Neither is it apparent that the granting of concessions to foreign investors or failure to exercise bargaining power must involve complicity. Political actors may inadvertently cause outcomes which favour foreign investors, for example if bureaucratic or political rivalries weaken a minerals department or other agency responsible for monitoring corporate activity or formulating mineral policy. It would be incorrect to say that complicity (which implies an element of deliberateness) occurs in such cases, and it should be noted that public pressure for a change in the actor's behaviour might not materialise, as no clear link might be perceived between that behaviour and the granting of the concession. In addition, a host government's ability to take advantage of improvements in its bargaining power may be constrained in important ways by political and ideological norms held by its constituents. For instance, a popularly-held belief in the 'sanctity' of private property may render unacceptable certain policy initiatives towards foreign investors.

Another issue involves the nature of the concession granted to foreign investors. Moran is concerned with the support of domestic groups for taxation and pricing policies which represent concessions to foreign investors, concessions which can of course be withdrawn almost immediately. If, on the other hand, the concession involves a reduction in host country bargaining power through dissipation of expertise, it may take a long time to reverse it.[53] In other words, in addition to the possibility that a country will not utilise the bargaining potential presented by progress along the learning curve, there is also the possibility that it will, so to speak, slip back along that curve.

This discussion raises a question in relation to Moran's concept of 'exploitation'. In his view, this results solely from the deliberate actions of domestic political groups. But there seems no a priori reason to accept Moran's statement that 'asymmetry of power per se between host countries and foreign investors [is not] a basis for "exploitation"'. Whatever the cause of an asymmetry of power (complicity, adherence to political norms, occurrence of 'pay-offs' elsewhere in the political system), if a foreign investor takes advantage of its existence he is exploiting that situation. This statement is descriptive, whereas Moran's concept of exploitation presumably has a prescriptive element in that complicity by domestic elites infers politically immoral behaviour on their part. It is descriptive because international mining companies will by their nature take advantage of a situation which enables them to increase profits and/or control over whatever time-period they employ, and one cannot rationally expect them to behave otherwise. But it is an important statement, in that identification of 'exploitation' with complicity can result in failure to recognise the relevance of other factors crucial to foreign investor-host country relations.

Two other general points arise. First, given that political norms favourable to foreign investors may retain their force over long periods of time, given that the actions of domestic political groups may not be linked in the public mind with the concessions which those actions give rise to, and given that it may be difficult to reverse certain concessions once they are granted, there seems no apparent reason why exploitative relationships between host countries and foreign investors should not continue on a basis which is, to all intents and purposes, permanent. This situation may be reflected in a failure either to overcome host country ignorance regarding mineral industries or foreign investor behaviour, or to take advantage of improvements in bargaining potential. To state such a possibility does not involve acceptance of a static 'neo-colonialist' viewpoint; it simply accepts that change in host country-foreign investor relations may take place to varying degrees and in different directions in different situations.

Second, it is not obviously true that a host country's failure to progress along a learning curve must imply a failure to maximise returns from foreign-financed mineral development. The host country may reach a point on the curve which allows it to devise a system of rules governing foreign investment so that the difference between the benefits it currently receives and those which could be obtained by further progress along the curve are less than the costs of making that progress. If those rules are permanent and generally applicable, yet flexible enough to take account of unexpected outcomes, then even limited conflict of the type envisaged by Mikesell may be avoided. In other words, the host country may be able to negate the effects of short-term shifts in bargaining power usually associated with major resource projects.

This of course assumes that the interests of the two parties are fundamentally reconcilable. Moran clearly assumes the opposite is the case,[54] an assumption stated more explicitly in Norman Girvan's studies of foreign mining investment in South America and the Carribbean.[55] Girvan believes that such investment results in a clash of

> two economic systems with totally different objectives and needs ... on the one hand the vertically-integrated multinational corporations ... on the other hand the national economy ... conflict between the corporate and national economies over the use of the same resource is therefore inherent and inevitable.[56]

Much more is involved here than a dispute over the division of economic rents, according to Girvan. The production and investment decisions of the foreign subsidiary will be determined by the global profit-maximising strategy of the parent company. The dictates of this strategy may coincide with the host country's interests at certain stages, but sooner or later shifts in relative production and other costs or the parent company's diversification policies will result in production and investment decisions inimical to the host country's interests.[57] Further, the host country will want to maximise linkages from mineral development by maximising employment, use of domestic sources of goods and services, and domestic processing of minerals. The foreign subsidiary will wish to minimise labour costs and will apply technology developed in its home country to do so.[58] It will tend to purchase goods and services from other subsidiaries within the group or through centralised purchasing agencies located in the home country, militating against development of domestic linkages. For various reasons, processing will also be largely carried on outside the host country.[59]

Even if host country governments capture a high proportion of economic rents, Girvan believes that attempts to foster growth in agriculture and manufacturing through their expenditure are unlikely to succeed because of the impact of foreign mining projects on the national economy. First, mining creates external diseconomies in relation to other sectors, for example by absorbing domestic resources and by raising wage rates. Second, the development of linkages from mining may well offer the best prospects for growth in other sectors, but of course these linkages do not materialise. Third, mineral development influences consumption patterns, encouraging consumption of expensive consumer durables which are difficult to produce locally, but which can easily be imported because of the abundant foreign exchange provided by mining itself.[60]

In Girvan's view, the only way in which a host country can avoid this situation is to secure total control over its mining industry, eliminating foreign investment and foreign influence altogether.[61]

The concern of authors such as Raymond Mikesell with 'joint-maximising strategies' reflects a belief that drastic action of this kind is not necessary in defence of a host country's interests, and indeed will have positively harmful effects. Though their interests will certainly clash on occasion, a fundamental conflict of interest does not exist between the foreign investor and the host country; both can be better off than in a situation in which foreign investment does not occur. Underlying this approach is the belief that the economic and other effects of

foreign mining investment are generally favourable,
or that appropriate host government policies can
render them favourable.[62]

Thus we are faced with a number of models which
posit certain conclusions regarding both the impact
of foreign investment and the nature of the foreign
investor-host country relationship. Two assume that
impact to be unavoidably adverse; a fundamental
conflict of interest consequently exists between
foreign investor and host country. One model predicts that this conflict will inevitably lead to
withdrawal of foreign investment, the other that
exploitative relationships may persist for extended
periods of time. Two other models assume that the
interests of both parties are fundamentally
compatible. One predicts that short-term shifts in
bargaining power will result in limited and
temporary conflict, the other suggests that such
conflict can be avoided.

It is of course unlikely that a limited number
of case studies will conclusively establish the
validity or otherwise of general propositions of
this kind. However, they may well allow judgements
as to individual aspects of each, and permit some
conclusions regarding their comparative usefulness
in analysing and explaining empirically-observed
phenomena. It is against this background that I turn
to an examination of the role of foreign investment
in mineral development in the case study countries.

Notes

1. If this were the case, the question would arise
as to why domestic investors should not have undertaken mineral development in any case. This question
might be dealt with in terms of the ability of
multinational mining corporations to pay a higher
price for mineral resources, and/or in terms of a
competitive edge possessed by such companies in the
discovery of minerals.
2. For example, during the 1960s and early 1970s a
foreign investor explored for, developed and brought
into production at a cost of A$400 million a massive
low-grade copper deposit on Bougainville Island in
Papua New Guinea. Domestic private investment on
such a scale was inconceivable: the necessary
resources, capital or human, simply did not exist.
Neither was there any possibility that the
Australian Colonial Administration would establish
and finance a Papua New Guinea State Mining Company
to develop the deposit, regardless of whether

foreign investors were willing to undertake the
project or not, as is evident from the
Administration's policy statements and actions at
the time (see Chapter Eight).
3. This approach characterises many socialist
critiques of foreign mining investment. See, for
example, Resources Study Group, Navan and Irish
Mining: Documentation of an £850,000,000 Robbery
(Dublin, 1972).
4. For a survey of some quantitative techniques used
during recent decades to analyse the economic
effects of foreign investments, see H.C. Bos, M.
Sanders and C. Secchi, Private Foreign Investment in
Developing Countries: A Quantitative Study on the
Evaluation of the Macro-Economic Effects (D. Reidel
Publishing Company, Dordrectht, 1974), especially
Part 1.
5. See, for instance, ibid., and G.D.A. MacDougall,
'The Benefits and Costs of Private Investment from
Abroad: A Theoretical Approach', Economic Record,
March 1960, pp. 13-35.
6. Bos et al, Private Foreign Investment, p. 376.
7. This applies equally to more traditional quantit-
ative methods and to recent attempts to establish
general statistical relationships between occurrence
of foreign investment and indicators of economic
growth (e.g. changes in GNP per capita). For some
examples of the latter, see V. Bornschier et al,
'Cross-national Evidence of the Effects of Foreign
Investment and Aid on Economic Growth and
Inequality', American Journal of Sociology, Vol. 84
(November 1978), pp. 651-83; R.W. Jackman,
'Dependence on Foreign Investment and Economic
Growth in the Third World', World Politics, Vol.
xxxiv, No 2 (January 1982), pp. 175-96.
8. See, for example, Bos et al, Private Foreign
Investment, Part II, Chapter II.
9. In each case study, the first-mentioned general
effect, generation of national incomes, is discussed
under four separate headings, Generation of
Government Revenue, Creation of Employment and Wage
Payments, Generation of Linkage Development, and Net
Foreign Exchange Inflow. The other general effects
are broken down in the manner which best facilitates
discussion of individual host countries or projects.
10. B. Smith and A.M. Ulph, 'Economic Principles and
Taxation of the Mining Industry: An Introductory
Survey', Paper Presented to the Workshop on Mining
Industry Taxation, Australian National University,
18-19 November 1976, p. 26.
11. For a fuller discussion of these factors, see

ibid., pp. 23-5.

12. In theory, the host country could confiscate 100 per cent of economic rents without affecting mineral development, but this would remove the incentive for operational efficiency, as after-tax profits would remain unchanged for a wide range of production costs.

13. Taxes are not neutral when they apply to that portion of revenue which is required to keep a factor (e.g. labour, capital) in a certain use. Where non-neutral taxes apply, resources will be allocated on criteria other than efficiency (which is assumed to be the basis of allocation in a 'free market' economy). Thus, a tax system should be designed so as to draw as high a proportion of revenue as possible from taxes which are neutral, that is taxes imposed on profits in excess of those required to keep a factor in a certain use. By definition, taxation of economic rents falls into this category.

14. Unit royalties are charged as fixed sums on each unit of production, ad valorem royalties as a percentage of the value of production.

15. The relationship between the variance in possible outcomes and expected rates of return is discussed by Ross Garnaut, 'Resource Trade and the Development Process in Developing Countries', in L.B. Krause and H. Patrick, Mineral Resources in the Pacific Area (Federal Reserve Bank of San Francisco, 1978), p. 146.

16. The same is often true where it is suspected that licence, patent or service fees paid by a subsidiary to a parent company are being utilised to transfer profits. For a discussion of this practice, see C. Brundenius, 'The Anatomy of Imperialism: The Case of Multinational Mining Companies in Peru', Journal of Peace Research, 3, 1972, p. 202.

17. The 'engine of growth' argument is summarised by A. Seidman, Natural Resources and National Welfare: The Case of Copper (Praeger, New York, 1975), pp. 11-12. For a general discussion of the role of linkages in economic development, see A.O. Hirschman, The Strategy of Economic Development (Yale University Press, 1958), Ch. 6.

18. The classic exposition of this view is that by H.W. Singer, 'U.S. Investment in Underdeveloped Areas: The Distribution of Gains Between Investing and Borrowing Countries', American Economic Review, Papers and Proceedings, Vol. XL (May 1950), pp. 473-85. For a more recent discussion of the way in which multinational corporations are integrated into the

international corporate economy, see N. Girvan, 'Multinational Corporations and Dependent-Underdevelopment in Mineral Export Economics', Social and Economic Studies, University of the West Indies, Vol. 19, No. 4 (December 1970), pp. 490-526. G. Lanning and M. Mueller argue that much recent mineral development in Africa has been of an 'enclave' nature. Africa Undermined: a history of the mining companies and the underdevelopment of Africa (Penguin Books, 1979).
19. See, for example, J. Grunwald, 'Foreign Private Investment: The Challenge of Latin American Nationalism', Virginia Journal of International Law, Vol. 11:2, 1971, pp. 235-6, 245.
20. B.L. Johns, 'Foreign Investment in Australia's Natural Resources', in J.A. Sinden (ed.), The Natural Resources of Australia: Prospects and problems for development (Angus and Robertson, 1972), pp. 297-8.
21. See, for example, J.B. Quinn, 'Technology Transfers by Multinational Companies', Harvard Business Review, November-December 1969, pp. 147-61.
22. The question arises as to why profit maximising firms would not use technologies which took advantage of domestic factor endowments, since they could presumably reduce production costs by doing so. The explanation might lie in the absence of appropriate equipment or techniques (reflecting the concentration of research and development in the developed countries), or in the comparatively low cost of capital within the economy of the mutinational corporation itself. N. Girvan, 'Multinational Corporations and Dependent-Underdevelopment', pp. 315-6.
23. Ibid.; P.E. Evans, 'National Autonomy and Economic Development: Critical Perspectives on Multinational Corporations in Poor Countries', International Organization, Vol. XXV, No. 3 (Summer 1971), pp. 682-3.
24. Where infrastructure is particular to mining operations, its provision cannot be regarded as a benefit to the host country in addition to the direct economic impact of mining itself.
25. R. Vernon, Sovereignty at Bay: The Multinational Spread of U.S. Enterprise (Basic Books, New York, 1971), p. 171; Y. Aharoni, The Foreign Investment Decision Process (Harvard University Press, Boston, 1966), pp. 152, 166-7.
26. Quoted in A. Mack, 'Theories of Imperialism: The European Perspective', Journal of Conflict Resolution, Vol. XVIII, No. 3 (September 1974), p.

519.

27. See, for example, L.T. Wells Jr, 'The Multinational Business Enterprise: What Kind of International Organization?', International Organization, Vol. XXV, No. 3 (Summer 1971), pp. 447-8; and S. Hymer, 'The Efficiency (Contradictions) of Multinational Corporations', American Economic Review, Papers and Proceedings, May 1970, p. 447.

28. This point is discussed in detail by Wells, 'The Multinational Business Enterprise'.

29. See, for example, R.F. Mikesell, Foreign Investment in the Petroleum and Mineral Industries: Case Studies of Investor-Host Country Relations (Johns Hopkins Press, Baltimore, 1971), Ch.2; R.F. Mikesell, Foreign Investment in Copper Mining: Case Studies of Mines in Peru and Papua New Guinea (Johns Hopkins Press, Baltimore, 1975), pp.24-33; Vernon, Sovereignty at Bay, Ch.2; D. Smith and L. Wells, Negotiating Third World Mineral Agreements (Ballinger Publishing Co., Cambridge, Mass.), Ch.1.

30. Mikesell, Foreign Investment in Copper Mining, p. 31.

31. Sovereignty at Bay, pp.40-53; 'Foreign Trade and Foreign Investment: Hard Choices for Developing Countries, Foreign Trade Review, Vol. 5, No. 4 (January-March 1971), pp. 504-5.

32. Mikesell, Foreign Investment in Copper Mining, pp. 32-3.

33. In Mikesell's words, 'the total rent from the production of the resources is not independent of the shares' (Foreign Investment in Mineral Industries, p. 40). In other words, should a host country government demand too high a share of the rent, total rent is likely to decline as a result of lower investment rates or other reactions on the part of the foreign investor.

34. Foreign Investment in Copper Mining, p. 29.

35. Foreign Investment in Mineral Industries, pp. 43, 49, 51-4.

36. T.H. Moran, Multinational Corporations and the Politics of Dependence: Copper in Chile (Princeton University Press, 1974); 'The Theory of International Exploitation in Large Natural Resource Investments', in S.J. Rosen and J.R. Kurth (eds.), Testing Theories of Economic Imperialism (Lexington Books, 1974), pp. 163-81; 'New Deal or Raw Deal in Raw Materials', Foreign Policy, No. 5 (Winter 1971-72), pp. 119-34.

37. Multinational Corporations and the Politics of Dependence, p. 93.

38. Ibid., pp. 158-62.
39. So in fact does the first: reduction in uncertainty involves a learning process through which the host country accumulates information regarding its own geological and other characteristics. However, here I am following Moran's typology, and he treats the two processes as separate developments. Multinational Corporations and the Politics of Dependence, pp. 162-3.
40. Ibid., pp. 163-6.
41. Ibid., pp. 8, 154.
42. Ibid., pp. 157, 222.
43. Ibid., pp. 157, 222, 224.
44. Ibid., p. 169.
45. Ibid., Ch. 6; 'The Theory of Exploitation', pp. 171-7.
46. Multinational Corporations and the Politics of Dependence, p. 156.
47. Documented cases of bribery of local elites by foreign investors are mentioned in F.B. Weinstein, 'Multinational Corporations and the Third World: the case of Japan and South-east Asia', International Organization, Vol. 30, No. 3 (Summer 1976), pp. 397-9; 'The U.S. calls a halt to multinational bashing', National Times, 16-21 May 1977.
48. See, for example, A.G. Frank, Capitalism and Underdevelopment in Latin America (Monthly Review Press, New York, 1967), pp. 115-9.
49. Multinational Corporations and the Politics of Dependence, pp. 217-8.
50. Ibid., pp. 156, 197-215.
51. Ibid., pp. 164-8.
52. Writing in 1974, Moran stated that 'no military dictatorship, no matter how suppressive of democratic rights, can reverse the movement towards Chilean control of basic natural resources for long' (Ibid., p. 248). The validity of this statement may well rest on the definition of 'for long', but at a time when private multinational mining corporations are again being assigned a major role in the development of Chile's mineral resources its validity can certainly not be taken for granted.
53. Technical and professional staff with the necessary experience are hard to find and, as they represent a relatively small group among whom communication is at a high level, adverse comments regarding a particular country spread quickly and can retard attempts to recruit replacement personnel.
54. This assumption is reflected in Moran's belief in the inevitability of fundamental conflict between

foreign investor and host country.
55. 'Multinational Corporations and Dependent-Under-development'; _Copper in Chile: A Study in Conflict between Corporate and National Economy_ (Institute of Social and Economic Research, University of the West Indies, 1972). This assumption also underlies some more recent studies of foreign mining investment; see, for example, Lanning and Mueller, _Africa Under-mined_.
56. _Copper in Chile_, p. 3.
57. Ibid., pp. 61-2; 'Multinational Corporations and Dependent-Underdevelopment', p. 504.
58. _Copper in Chile_, p. 16.
59. Ibid., p. 50; 'Multinational Corporations and Dependent-Underdevelopment', pp. 513-5.
60. 'Multinational Corporations and Dependent-Under-development', pp. 520-3.
61. Ibid., p. 526; _Copper in Chile_, pp. 61-2.
62. Mikesell, _Foreign Investment in Mineral Industries_, pp. 24-5.

Chapter Two

FOREIGN INVESTMENT AND THE DEVELOPMENT OF THE IRISH
NON-FERROUS MINING INDUSTRY, 1956-1977.

The Republic of Ireland is a country of some 26,000
square miles with a population of approximately 3.1
million. The twenty-six counties which comprise the
Republic became a separate political entity in 1922.
At Independence, the Irish Free State, as it was
then known,[1] was a predominantly agricultural
country. This situation resulted in large measure
from the absence of local supplies of key minerals
(especially coal and iron ore), from the country's
colonial status as a specialised agricultural
producer for the British market, and from the
exclusion of the most heavily-industrialised section
of Ireland (the North East) from the Irish Free
State. Over the next fifty years, its occupational
profile changed considerably, so that by 1974 22.6
per cent of the population were engaged in
agriculture and 29.1 per cent in industry, versus 50
per cent and 12.5 per cent respectively in 1922.
This change was heavily concentrated over the period
1946-74 and was associated with a reorientation of
industrial development strategy towards
encouragement of export-oriented foreign-financed
projects.
    Since 1922, the Republic of Ireland has been
governed under a Westminster-type parliamentary
system, and its Dail (Parliament) has been dominated
by one or other of the two largest political
parties, Fianna Fail and Fine Gael. In general,
economic policy has been oriented towards
encouragement of private initiative, though
government action (for example, generation of
electricity from native fuel resources) has been
taken where such initiative was lacking.
    As mentioned above, Ireland has not enjoyed an
abundance of mineral wealth in the past. Over the

38

period 1960-80, this situation changed dramatically. By 1978, the Republic of Ireland had become the world's fifth largest zinc producer, its third largest barytes producer, and a significant producer of lead and silver. This chapter explains why and how this change came about, and examines the economic and political impact of foreign-financed mineral development in Ireland.

Historical Introduction

During the period 1820-80, copper, lead and zinc mining occurred on a large scale in Ireland, but by the late 1870s the industry was in decline, and by 1885 mining of non-ferrous metals had almost entirely ceased. After Independence, interest in the development of indigenous mineral resources revived. However, various factors militated against pursuit of a more vigorous exploration campaign in the years prior to 1940. Uncertainty existed over the ownership of minerals in many areas, a situation which was only remedied by the introduction of the Mines and Minerals Development Act in 1940. The absence of currently operating mines added to this uncertainty, reducing Ireland's attractiveness to private mining concerns. The Irish government either could not afford to, or did not wish to, become directly involved in exploration on a large scale. Speaking in 1936, the then Minister for Industry and Commerce, Mr Lemass, made his government's views quite clear. 'The development of mineral resources is a matter primarily for private enterprise and I am not prepared under present circumstances to undertake the acquisition and exploitation on behalf of the State of mines...'[2]. As private enterprise was not enthusiastic in pursuing its allocated task, significant mineral development was unlikely to occur.

Raw material shortages caused by the outbreak of war in 1939 led to a change in government policy. As part of a drive to fully utilise indigenous resources, a state mining company, Mianrai Teoranta, was established and given responsibility for the development of metallic mineral reserves. In 1947 Mianrai Teoranta commenced detailed exploration near Avoca, Co. Wicklow, and by 1955 the Company claimed the existence of some 13 million tons of ore, with a copper content over one per cent. Mianrai Teoranta was not allocated the task of developing the deposit; negotiations were undertaken with Consolidated Mogul Mines of Canada, and in 1956 a mining lease was granted to Mogul's wholly-owned subsidiary, St Patrick's Copper Mines Limited. The government's attitude was that state involvement in industry was only justified where private enterprise failed to show any interest; private interests were willing to develop the Avoca deposits, and consequently there was no need for further involvement by Mianrai Teoranta.[3]

## The Avoca Development 1956-1962

St Patrick's Copper Mines Limited was granted a lease to mine the Avoca deposits on the following terms. The Company was to spend stipulated amounts on exploration and development of the area; if the results warranted commercial development, St Patrick's would construct a concentrator and other essential plant at a cost of approximately £2 million. It undertook to employ 500 men, and to sell the products of the mine at the best price obtainable (defined by an elaborate formula included in the lease). The Company would reimburse the State for Mianrai Teoranta's exploration expenditure.

The taxes to be borne by St Patrick's were not heavy. It would pay a royalty of 4 per cent on profits up to £350,000, rising to 9 per cent on profits in excess of £1,750,000. In February 1956, taxation legislation was introduced which had an important bearing for the Company. The Finance (Profits of Certain Mines) (Temporary Relief from Taxation) Bill proposed to grant total exemption from corporation tax for a period of four years, and 50 per cent exemption for a further four years, to Irish-registered mining companies commencing operations within three years of the enactment of the Bill.

During the Dail debate on the Bill, its provisions were the subject of harsh criticism by the opposition Fianna Fail party. The party's spokesman for Industry and Commerce, Mr Lemass, later to become Taoiseach (Prime Minister), claimed that the Bill was 'designed to encourage raids upon Irish mineral resources by foreign financial interests with the idea of getting as much profit out of their operations as quickly as possible'. He estimated that Mogul would make a net profit of approximately £13 million over the life of the mine, while Ireland's financial returns would be negligible. Other prominent Fianna Fail members, including the future Irish President, Mr De Valera, were equally critical, but despite their opposition the Bill was passed without major amendment.[4]

Mogul carried out its investigations, decided to proceed with the project, and by late 1958 had installed the concentrator and other equipment. During the development stage, St. Patrick's fully utilised the £1.76 million raised on its behalf by Mogul in Canada. Further funds were needed to complete the project, but Mogul was unable to obtain these in Canada, and it requested the Irish government to guarantee a loan for the amount required. The (now Fianna Fail) government agreed, having received a favourable assessment of the project's prospects from a reputable mining investment house. A loan of £1,368,000 was obtained from the Irish Life Assurance Company.

Soon after production commenced, St Patrick's ran into difficulties. Its management made a number of poor technical decisions[5]; these resulted in lower-than-expected ore grades and reduced production which, combined with depressed copper prices, created serious financial difficulties for St Patrick's. It arranged a second State-guaranteed loan of £550,00, and obtained a government grant of £240,000 to help finance further development work, on Mogul's assurance that this would permit production of ore of an economic copper content. This assurance was apparently based on a projected rise in world copper prices, which failed to materialise. Mogul did not have, or was unwilling to commit, the resources to tide St Patrick's over an extended period of low copper prices. St Patrick's went into receivership in May 1962, and, despite the injection of another £250,000 by the Irish government over the following months, operations had ceased by September 1962. St Patrick's defaulted on its loans, which became the liability of the Irish government.

The political repercussions of the Avoca failure for the Fianna Fail government were immediate and serious. Mr Lemass, now Taoiseach, and Mr DeValera, now President, were publicly ridiculed for their earlier claims concerning the project's likely profitability. More serious were the accusations that government support for St Patrick's Copper Mines Limited represented a wasteful use of public monies.[6] It mattered little that the Company's failure was primarily due to the vagaries of the world copper market[7] and to the technical and investment decisions made by a foreign corporation. Future Irish governments would be very reluctant to permit state financing of mineral development, and the Fianna Fail politicians who were to dominate

those governments would not be disposed to believing
that mining companies operating in Ireland were
enjoying excessive profits.

The Avoca development had a major impact on the
direction of mineral policy during the following
decade. The period 1956 to 1962 saw an end to direct
state involvement in the mining industry, the
initiation of a policy of affording financial
incentives to mining companies, a disastrous outcome
to the State's first major financial involvement in
the mining industry, and an embarrassing experience
for the Fianna Fail politicians who were to rule
Ireland until 1973. As will become apparent, these
events were to cast a very long shadow indeed.

The Growth of Mineral Exploration and Development,
1958-1968

An important side effect of the Avoca development
was that it brought Ireland to the attention of a
number of mainly Canadian-based mining companies. In
the years after 1956, exploration activity
increased. Initially, attention was focused on
locations where mining had previously taken place,
but towards the end of the 1950s a number of
companies started to apply geochemical techniques
which allowed prospecting of 'virgin' areas.[8] One of
these was Irish Base Metals Limited (IBM), a wholly-
owned subsidiary of a Canadian-based company,
Northgate Exploration Ltd. Northgate had been set up
in 1958 by three Irishmen who had emigrated to
Canada in the late 1940s, Pat Hughes, Michael
McCarthy and Matt Gilroy, and had there gained
experience of the mining industry. Capital was
raised by selling some 80 per cent of Northgate's
stock to Canadian investors, while the three
founders of the Company themselves retained
significant shareholdings.

In September 1961 IBM undertook geochemical
soil and water sampling in the Tynagh area of Co.
Galway. The results were promising, and a drilling
programme was initiated. This revealed the existence
of substantial reserves of lead ore, which also
contained significant amounts of silver and copper.

From 1962 to 1964, preparations were undertaken
for the development of the Tynagh deposit. The
Minister for Industry and Commerce issued an order,
under the 1940 Mines and Minerals Development Act,
taking possession on behalf of the state of all
minerals in private ownership or concerning which
ownership was doubtful, to ensure the orderly

working of the deposit and to guard against obstructionist behaviour by private individuals. In December 1962 IBM was granted a 21-year mining lease, which provided for royalty payments on a graduated scale to a maximum of 9 per cent on profits in excess of £1.75 million per annum.[9] As the provisions of the 1956 taxation legislation had been extended by the 1959 Finance Act, the operating company would be exempt from corporation and income tax for a period of four years, and enjoy 50 per cent exemption for a further four. Major development work commenced at Tynagh in 1964, and by late 1965 the mine was in operation.

The Tynagh discovery brought Ireland to the attention of the international mining industry, and led to a spectacular increase in mineral exploration and development. In 1963 Mogul Mines Limited discoved a major zinc/lead deposit at Silvermines, Co. Tipperary. Production commenced in May 1968, and Silvermines became Europe's largest underground producer of zinc and lead. In late 1967 Gortrum Mines Limited commenced production of copper and mercury at Gortrum, Co. Tipperary, and in early 1971 the Avoca copper mine was reopened by Avoca Mines Limited, a subsidiary of a Canadian Company, Discovery Mines. A large baryte-mining operation was established near Silvermines by Magnet Cove Barium of the United States in 1963, and in 1969 a subsidiary of the Pfizer Corporation commenced mining of dolomite at Bennetsbridge, Co. Kilkenney.

Two factors were particularly important in facilitating this dramatic increase in exploration and mining activity. The first was the existence of workable, efficient legislation relating to mineral ownership. A high proportion of mineral rights was owned by the state, while the Mines and Minerals Development Act 1940 empowered the Minister for Industry and Commerce to compulsorily acquire any rights still in private hands, should this prove necessary. This meant that a company could 'acquire directly from the Government the right to explore a given area with the assurances that it can get a mining lease ... and that a present owner could not block a broad development', in contrast to the situation in Britain, for example, where a company might 'have to deal with a large landowner, several small farmers, numerous lots owned by retired men or their widows, perhaps a land developer and the estate of someone who emigrated to Australia thirty years before and died intestate.'[10] Under the British system a company might spend two years in

obtaining satisfactory title to a particular area.
In Ireland, this would take literally a few days,
and in addition the company would know that, if its
interest was justified, problems would not arise in
obtaining permission to proceed with development.
   The availability of modern exploration
techniques, particularly geochemical ones, was
equally crucial. Geochemical testing, or the
analysis of soil and water samples for traces of
metal, permitted collection of data over large areas
devoid of surface indications of mineralisation, and
thus facilitated the pin-pointing of promising
locations for further investigation. This technique
was crucial to the discovery of the Tynagh deposit,
a discovery which provided the fundamental
prerequisite for a vigorous exploration programme,
i.e. favourably altered perceptions as to the
geological likelihood of discovering a major mineral
deposit in Ireland.
   Another important factor was the 1956 taxation
legislation, as amended in 1959.[11] While the
financial inducements offered by this legislation
were not as significant as the factors discussed
above, they did have a triggering effect. They
achieved publicity in international mining circles,
and offered the prospect of a valuable concession to
any company fortunate enough to discover a
commercially viable deposit. Such concessions would
have had little effect, however, had mining
companies felt that the possibility of discovering
worthwhile deposits was remote, or that problems
concerning land or mineral ownership would have made
it difficult or impossible to develop deposits which
were found. This legislation was also important in
that it expressed an attitude on the part of the
Irish government - a welcoming attitude, a
determination to facilitate foreign mining companies
in exploring for and developing minerals in Ireland.
   For these reasons, Ireland became an attractive
location for mining investment. The companies which
became involved in mineral development were
predominantly foreign owned and financed.[12] Local
investors were not denied the opportunity to become
involved, but rather chose not to.[13] Mining was
perceived in Ireland as a high-risk activity (a
perception heigthened by the failure of the Avoca
development), and Irish investors had long fought
shy of areas where the level of risk seemed high.
Ireland also suffered from a scarcity of the
entrepreneurial and managerial skills so essential
for successful initiatives in the mining sector.[14]

Thus the field was left to foreign investors, and it seems certain that had they not found Ireland an attractive investment location, mineral development would not have occurred in the 1960s.

## The Twenty-year Tax Exemption

In 1967 the Fianna Fail government introduced a twenty-year tax exemption for foreign mining companies operating in Ireland. This exemption, announced in the Budget Address on 11 April, replaced the existing provisions which allowed four years of total exemption and a further four years exemption on 50 per cent of profits. As none of the mines then known to exist had a life expectancy of more than twenty years, the Irish government was effectively denying itself the possibility of obtaining significant revenue from the mining industry.[15]

By 1967, the extent of uncertainty associated with mineral development in Ireland had diminished greatly. Five major mineral deposits had been discovered during the previous decade (Tynagh, two deposits at Silvermines, Gortrum, and Bennetsbridge) and Tynagh and the baryte operation at Silvermines were already operating profitably. Considerably more information was available concerning geology and operating conditions in Ireland than had been the case a decade previously. Ireland had also progressed along a 'learning curve' to some extent, in that members of the Department of Industry and Commerce and of the Geological Survey Office were developing expertise in relevant areas. These changes might be expected to have placed the Irish government in a stronger bargaining position, allowing it to impose more stringent terms on foreign mining concerns. It did the opposite, introducing terms very much more generous than those which had previously applied. What explains this action?

When announcing the government's decision in his Budget speech, the Minister for Finance, Mr Haughey, gave little indication of what the answer might be. He stated:

> I am told that a great future expansion in exploration and development is possible and that additional tax incentives can generate a large volume of outside investment ... Many different allowances and incentives have been suggested but, instead of bringing in rather

complicated new provisions, I have come down in
favour of the simple decision to substitute for
the existing reliefs a twenty-year period of
complete exemption.[16]

Mr Haughey's justification of the measure
consisted in the claim, therefore, that the
concession would lead to greater foreign investment
in the industry and that it was simple in nature.
The first argument might have held some force had
the concessions been applicable only to mines
discovered after the legislation came into effect,
but it is difficult to understand why additional
concessions to existing, profitable, mines were
needed to encourage future exploration. In addition,
the government had received clear evidence in the
previous year that its existing tax concessions were
successful in encouraging increased foreign
investment (see below).

In fact it is apparent that no specific
consideration was given to the question of what type
of measure would best serve Ireland's interests. The
head of the Mines and Minerals Section of the
Department of Industry and Commerce, whose staff
would have been involved in such an exercise, was
not even consulted on the legislation.[17] The Irish
government clearly did not utilise the expertise
made available by progress along a 'learning curve'
by its officials. For example, it also failed to
consult the then   Technical Adviser to the Minister
for Industry and Commerce (an employee of the
Geological Survey Office), who in fact believed that
a more generous exemption was totally unnecessary,
given the level to which the mining industry had by
then developed.[18]

The complexity of taxation provisions is hardly
a sufficient justification for not imposing any
taxes. As one commentator noted wistfully:

It is my sincere regret that neither Mr Haughey
nor any of his successors in the Ministry of
Finance ever found the laws on income tax
equally complicated. There are many wage and
salary earners who would crave for the simpli-
city of a twenty year exemption.[19]

Another circumstance which requires explanation
is the failure of opposition party spokesmen to
offer any criticism of the measure. Throughout the
debate on the Budget Address, or during the passage
of the legislation through the Dail, not a single

voice was raised to criticise the granting of so generous a concession.

A credible explanation for the Irish government's behaviour has not been advanced to date. Critics of the legislation have claimed that Irish politicians served the interests of international mining companies without any concern for the national welfare, and that the twenty-year tax exemption was simply the most blatant example of this tendency. However, these critics have not explained _why_ Irish politicians should behave in such a manner.[20] I believe a 'pay-off' was made by Irish politicians to foreign mining companies, but that this occurred precisely because those politicians were pursuing the national interest as they perceived it. This explanation requires some understanding of developments in Irish economic policy over the decades to 1967.

Since Independence, Ireland's most pressing economic difficulties have related to continuously high levels of unemployment, associated with a high level of emigration, and, more immediately, to problems in balancing its international payments. During the 1930s and 1940s, a policy of import-substitution aided by substantial tariff barriers was implemented in an attempt to deal with these problems. But this strategy enjoyed only limited success, particularly because the small size of the domestic market severely limited the opportunities for growth of import-substituting industries. Unemployment remained high, emigration in the early 1950s was at its highest since the 1880s, and in 1955 Ireland faced a balance of payments crisis of such magnitude that, as one commentator noted, 'it provoked doubts in some minds as to the viability of an independent Irish State'.[21] These circumstances led to the conviction that an alternative policy to protectionist import-substitution was required.

A major step towards a new approach was taken by the Coalition government in 1956 through certain provisions of the Finance Act 1956. These provided for partial remission of corporation and income taxes on profits received through new or increased export sales. But the new strategy was not explicitly formulated or implemented as a coherent policy until the return to power, in March 1957, of a Fianna Fail government. The basis of the new policy was to be the creation of mainly export-orientated industrial concerns financed by foreign investment. Foreign investment was to be encouraged by the provision to such concerns of generous taxation concessions, by a system of government grants to help foreign companies set up subsidiaries in Ireland, and by a range of other incentives. Effect was given to this policy through legislation introduced in 1957-9.

This strategy was designed to attack the fundamental problems facing the Irish economy, without interfering unduly with Irish-owned industries established under the earlier protectionist regime. By concentrating on exports, it would provide wider outlets for Irish industrial production, thus removing the limits placed on economic growth by the small size of the domestic market. Increased exports would alleviate the balance of payments problem, which preoccupied politicians after the crisis of 1955. Inflow of capital could also be expected to increase, with similar effect. Most important of all, opportunities for employment would not be restricted by the limits of import-substituting industries.

In effect, the Irish government was offering foreign investors a trade-off. It offered them a 'pay-off' in the prospect of tax-free, profitable investment in a 'favourable' business climate. In return, it expected to see additional employment and export receipts generated. Thus Irish politicians could credibly present themselves to their constituents as being capable of dealing with what were generally seen as the most serious economic problems facing Ireland. This was to be their 'pay-off'.

The Irish government's approach to the mining industry now becomes fully comprehensible. Foreign investors were prepared to develop an industry which would export 100 per cent of its output. In so doing, they would add to capital inflow, provide employment, not compete with any local industrial enterprise, and, unlike manufacturing concerns, not require government subsidisation through provision of grants or facilities. In fact, they would give a government pursuing the strategy outlined above all it could ask for. Thus it was to be expected that, like other foreign-financed industrial concerns, mining companies would receive a quid pro quo. In this instance, it took the form of a generous tax concession. An important element in this explanation is that Irish politicians assumed that foreign mining companies, like their counterparts in other sectors, did in fact require a 'pay-off' if they were to undertake (or to continue) mineral production in Ireland.

This interpretation offers an explanation for a number of the discrepancies mentioned earlier. First, it explains the failure of Irish governments to take advantage of favourable changes in their bargaining power. Utilisation of such changes would have been inconsistent with the perceived need to offer a 'pay-off' to foreign investors. Similarly, the expertise which had been acquired by members of the Department of Industry and Commerce and of the Geological Survey Office was not needed, and this explains why the relevant individuals were not consulted in 1967. Second, the failure of the Fianna Fail government to justify its decision on the twenty-year tax exemption in a credible manner is understandable, in that government ministers took for granted public acceptance of the wider strategy of which that decision formed a part. By 1967, this acceptance was almost universal among the Irish public and among politicians of all the major political parties,[22] which explains the failure of the opposition to criticise the tax exemption in parliment.

A number of important questions remain. First, was it in fact necessary for the Irish government to make a 'pay-off' in order to ensure that foreign investors would undertake mineral production in Ireland? Second, did Irish decision-makers adopt the correct strategy in pursuit of their policy aims? In other words, has foreign mining investment generated the benefits expected of it?

Foreign Investor-Host Country Relations in Ireland

In dealing with the first question, a distinction must be maintained between the concessions offered to foreign investors by the twenty-year tax exemption and those provided by the 1956 legislation. Much of the evidence available in 1967 suggested that the granting of a more generous concession to foreign mining companies was unnecessary. This was certainly true of those companies which had committed funds to mine development, and were already (or soon expected to be) earning substantial profits. It is possible that some additional concession was needed to encourage development of marginal mines, though the decision to develop Gortrum, a marginal deposit, was taken before 1967. It would however have been possible, as was done later,[23] to provide such a concession without foregoing revenue from profitable mines.

The question of investment in exploration is

more complex, but here too the available evidence indicated the efficacy of existing legislation. During the years 1963-65, exploration activity declined from a peak reached in the wake of the Tynagh discovery in 1961. In 1965, the qualification period for the 1956 tax legislation was extended by ten years to April 1976. This gave a significant boost to exploration, and in the following year the number of new companies becoming involved in exploration increased substantially (14 as opposed to six in 1965 and two in 1964), as did the number of prospecting licences held and the total area covered by such licences.[24] The decline in exploration activity in 1963-5, combined with this response to the extension of the 1956 legislation, indicate that it was still necessary to offer some financial incentive to attract foreign investment into mineral exploration of Ireland. It is of course possible that the concessions offered by the 1956 legislation could have been reduced to some extent without affecting the flow of investment.

Thus the evidence available in 1967 indicated that the granting of a twenty-year tax exemption was not necessary to the maintenance of mineral production, or to ensure continued investment in exploration. What lay behind the failure of the Fianna Fail government to correctly interpret the situation at the time? The immediate explanation is quite simple. Foreign mining companies claimed that they needed more generous tax concessions, and the government believed them. To understand this response, it is necessary to look more closely at the nature of relationships between foreign mining concerns and Irish politicians.

First, two aspects of the manner in which Irish politicians viewed foreign mining concerns must be emphasised. They clearly believed that Ireland should be grateful to foreign concerns which brought in their capital and expertise and thus facilitated mineral development. As one Deputy expressed it, 'We should welcome to this country anybody who is prepared to invest money here and employ people who would otherwise have to emigrate.'[25] A number of years later, another Deputy stated the view that, 'We should be grateful to people who are prepared to come in and invest their own money', while one of his colleagues felt compelled to 'pay a special tribute to the very progressive Canadians who have come here to help in the development of our mineral resources.'[26] There seems to have been no awareness that Ireland was engaged in an exchange relationship, that valuable mineral resources were being made available to foreign investors in return for their capital and expertise, and that Ireland should apply the bargaining power represented by mineral ownership to ensure maximum returns to itself from their exploitation. Rather, a 'begging bowl' attitude was predominant: Ireland was fortunate to receive foreign mining investment at all, and was consequently fortunate to receive any returns from development of its minerals.

The second point is that Irish politicians clearly believed that while foreign mining companies knew their business well, the Irish authorities knew very little about the mining industry. Consequently any intervention on their part might have adverse effects. This view was strengthened by frequent references to the role supposedly played by the government of the day in the Avoca debacle. From the the mid 1960s onwards, a number of prominent politicians frequently repeated the assertion that government intervention had been responsible for the failure of the Avoca mine.[27] As we have seen, the prime responsibility for that failure lay with the foreign investor concerned. Nevertheless, statements attributing blame to the government were not challenged in the Dail, reflecting a general acceptance of the viewpoint they expressed. Neither was it recognised that the issue of government involvement in mining operations was quite separate to that of its role in determining the financial and economic framework within which mining should take place.

Irish politicians, therefore, had scant regard for their own bargaining position or for the ability of Irish authorities to exercise expertise in the mining field. On the other hand, they held the expertise and knowledge possessed by foreign mining companies in very high regard. Their response to this situation was to accept as valid the mining industry's account of what it needed. They drew no distinction between what the industry needed and what it wanted, with the result that government policy was usually determined by what foreign mining companies asked for.

There is general agreement that the introduction of the twenty-year tax exemption followed a period during which foreign mining concerns had demanded more generous concessions.[28] The Irish government apparently decided to accept without question the claim that a greater incentive was required to ensure maintenance of production and continued investment in exploration. If the Minister of Industry and Commerce had consulted his departmental officials, they would have pointed out that what the mining companies wanted was not necessarily what was required to ensure their continued involvement. That he failed to do so is clear evidence of the automatic way in which government policy responded to the demands of foreign mining companies.

It seems likely that those same companies played a role in fostering and enforcing the attitudes discussed above. It is very difficult to present conclusive information concerning the impact of foreign investors on the outlook and attitudes of politicians. The latter will rarely admit to being influenced by the representatives of foreign corporate interests, while foreign investors are loath to acknowledge they possess influence over decision-makers in the host country. Yet it is not possible to ignore this issue simply because relevant documentary evidence is not available.

In the Irish case, it is quite clear that close relationships existed between various politically influential individuals and officials of foreign mining companies; the President of the Northgate group, for instance, was a personal friend of the Deputy Leader of the Labour Party, and of a number of senior Fianna Fail ministers. These relationships were probably strengthened by the fact that a number of family relations of senior politicians were employed by foreign mining companies. (This is not to suggest that any of the individuals involved were not qualified for the positions they held.) For example, the son of the Deputy Leader of the Labour Party became Secretary of Tara Mines Limited (an associated company of Northgate's), while the brother-in-law of a Minister for Industry and Commerce was also employed by Tara. Without engaging in speculation, it can be said that relationships of this type might have acted as a channel for ideas and opinions and represented a means by which influence could be exerted.

Did foreign mining investment generate the economic benefits expected of it? The following sections examine the economic impact of foreign-financed mining projects over the period 1964-77. In certain instances, data required for analysis of the non-ferrous mining sector as a whole could not be obtained, and in these cases the operations of the Northgate group (concerning which information was available) are analysed separately. The conclusions

derived from such an analysis are probably valid for
the sector as a whole, as the other projects
(Silvermines, Avoca) were similar to Tynagh and
Gortrum in the mining techniques employed, in the
materials produced, in the extent to which they were
foreign owned, and in being entirely financed by
foreign capital. At times use is made of financial
statistics relating to the Northgate group, and the
relevant data is presented in Table 2.1.

Generation of Government Revenue

Direct revenue accrued to the government in two
ways. Royalties were charged on net profits
throughout the period under review, on a rising
scale to a maximum of 10 per cent. Mining companies
were exempt from corporation tax until 1973,[29] after
which they became liable at the standard rate (then
50 per cent).
      Lack of data makes it difficult to accurately
assess taxation receipts. The Irish government does
not publish details of royalty payments by
individual companies, and industry sources are
frequently unreliable (see below). For the industry
as a whole, royalty payments to end 1977 were
£3,360,000. Official information regarding
corporation tax is also sparse, but accurate figures
can be calculated from the published accounts of
individual companies, and these are presented in
Table 2.2. The Irish Government received significant
amounts of corporation tax after the ending of the
industry's tax exemption in 1973, to the tune of
approximately £9 million in five years. A very high
proportion of this revenue was generated by Mogul of
Ireland's mine. The Northgate group contributed
little, because the profitable high-grade mining
period at the Tynagh mine was over by 1974, when
Northgate first became liable to taxation.
      Government revenue also accrues from employee
taxation.[30] On the basis of figures supplied by an
industry source, employee taxation has been
estimated at £12.5 million over the years 1964-77.
It was thus similar in scale to total direct
taxation over the same period (about £12 million).
This situation reflects the very low levels of
direct taxation applied prior to 1973, and is
unlikely to continue into the future.
      Little opportunity has existed for transfer
pricing in the non-ferrous mining industry, as the
operating companies sell their output on an 'arms

# Table 2.1 : Financial Statistics, Tynagh and Gortrum Mines, 1966 - 77 (C$000s)

| | 1966* | 1967* | 1968* | 1969* | 1970 | 1971 | 1972 | 1973 | 1974 | 1975 | 1976* | 1977* | Total** |
|---|---|---|---|---|---|---|---|---|---|---|---|---|---|
| Net Sales Revenue | 10,245 | 15,442 | 16,118 | 15,986 | 19,646 | 13,187 | 16,715 | 28,116 | 25,945 | 20,418 | 11,608 | 13,908 | 207,334 |
| Deduct Expenses as Follows: | | | | | | | | | | | | | |
| Operating Expenses*** | 3,387 | 4,804 | 4,736 | 4,293 | 8,318 | 8,669 | 9,303 | 12,395 | 14,314 | 16,374 | 8,790 | 10,660 | 106,143 |
| Government Royalty | 420 | 738 | 738 | 887 | 834 | 61 | 263 | 1,171 | 911 | 19 | 190 | 247 | 6,479 |
| Taxation Attributable to Mining Income | – | – | – | – | – | – | – | – | 2,686 | 51 | –[+] | –[+] | 2,737 |
| Depreciation | 967 | 1,142 | 1,399 | 2,112 | 2,677 | 2,431 | 2,431 | 2,797 | 3,083 | 2,357 | 1,806 | 1,599 | 23,435 |
| Amortization of Preproduction and Other Expenditur | 542 | 779 | 791 | 530 | 1,425 | 1,510 | 1,244 | 1,373 | 1,172 | 998 | 756 | 754 | 11,874 |
| Interest Payments | 1,257 | 614 | – | – | 338 | 303 | 317 | 249 | – | – | – | – | 3,078 |
| Exploration | – | 339 | 495 | 519 | 979 | 1,081 | 422 | 526 | 594 | 1,034 | 1,541 | 2,371 | 9,901 |
| NET INCOME (NORTHGATE GROUP) | 3,783 | 7,605 | 8,346 | 8,721 | 6,268 | (1,306) | 3,094 | 9,120 | 4,026 | 1,156 | (598) | (590) | 49,625 |
| - Attributable to Mining | 3,672 | 7,103 | 8,216 | 8,358 | 5,640 | (1,114) | 2,735 | 9,605 | 3,185 | 169 | (1,475) | (1,723) | 42,194 |
| Capital Expenditure ++ | 1,022 | 995 | 1,020 | 1,382 | 3,790 | 5,573 | 1,709 | 739 | 1,043 | 372 | 455 | 243 | 18,343 |
| Dividends Paid | – | – | 2,725 | 2,770 | 2,826 | 1,510 | – | – | 4,059 | 1,723 | – | 1,850 | 17,463 |

* For these years, the data applied to the Tynagh mine alone

** It should be stressed that the 'Total' columns in this and other similar tables presented below do not attempt to represent income streams from the projects involved.

*** These include operating costs at the mine and also administrative and general expenses

+ As no mining income was received in these years, it is assumed that no tax liability was incurred.

++ Capital expenditure consists of additions to fixed assets and outlays on the underground development at Tynagh.

Source: Northgate Exploration Limited, Annual Reports, 1966-1977.

Table 2.2 : Income Tax Liability of Mining Companies in Ireland, 1973 - 7 (£000s)

| | 1973 | 1974 | 1975 | 1976 | 1977 | Total |
|---|---|---|---|---|---|---|
| Northgate Group* | -** | 1,124 | 23 | - | - | 1,147 |
| Mogul of Ireland | 1,176 | 1,656 | 1,356 | 2,316 | 1,188+ | 7,692 |
| Avoca Mines Ltd. | 50++ | - | - | - | - | 50 |
| | | | | | | 8,899 |

\*    On income attributable to mining only.

\*\*    Northgate's mining income was subject to tax in 1973, but because of accumulated deductible allowances no liability was actually incurred.

\+    In 1977 Silvermines Ltd. did not provide information on Mogul of Ireland's tax liability (Annual Report, 1977, p.4). It is presumed that Mogul paid corporation tax at the prevailing rate of 45 per cent.

++    This liability did not lead to a tax payment.

Source:   Annual Reports, Northgate Exploration Limited, Silvermines Limited, various years; information supplied by Avoca Mines Limited.

length' basis,[31] and as none of their parent companies hold significant interests in overseas producers of mining inputs.[32] However, examination of Northgate's accounts over the period 1962-77 did reveal a number of discrepancies.

First, the Company's published figures overestimate both the liability of the Irish subsidiaries to royalty payments and the payments actually made. For example, the Tynagh lease called for royalty payments of £257,000 or C$662,000 in 1969,[33] while Northgate stated royalty payments to have been £345,000 or C$887,000. This divergence was not due to any time lag in collection of royalties, as it occurs throughout the period 1966-77. During these years, total stated royalty payments by the Irish subsidiaries amounted to 13 per cent of profits attributable to mining,[34] whereas the maximum rates applicable to the Tynagh and Gortrum mines were 9 and 10 per cent respectively. (A significant proportion of profits would have paid royalties at lower rates.) In addition, the royalty rates called for in the mining leases were not in fact applied. In 1968, for example, total royalty receipts by the Irish Exchequer amounted to £200,867. (A number of mining and quarrying projects other than Tynagh were also liable to royalty payments.) Yet Tynagh's liability in that year was £252,000.[35]

Northgate's motives for overestimating its royalty payments are somewhat of a mystery. The only explanation which presents itself is that the Company wished to exaggerate its contribution to the Irish economy for public relations reasons.

The second discrepancy relates to the application of amortization and depreciation allowances to Northgate's Irish operations. Expenditures associated with mine development and not allowable for depreciation under general tax provisions were amortized over the estimated life of the mine; items which were so allowable were depreciated over varying time periods.[36] It is evident that an element of double depreciation was involved in these procedures. The Tynagh mine was still in operation in 1977, and amortization and depreciation was consequently incomplete. Thus the sum of amortization and depreciation allowances should have been less than the sum of all relevant expenditures. In fact it was C$3.9 million more, indicating that double depreciation had been effected to an amount in excess of this sum.[37]

The result was that Northgate's taxable income,

and consequently government revenue, was diminished. The allowances claimed by Northgate would of course have been sanctioned by the taxing authority (the Department of Finance), and the question arises as to why Irish officials should condone loss of tax revenue. I return to this question at a later stage, and suggest a possible explanation (see p.77 below).

Generation of Employment and Wage Payments

Mining operations in Ireland are highly capital-intensive, and their employment-generating potential is consequently limited. For example, the average capital investment per job in overseas grant aided firms operating in Ireland was £7,850 in 1976; the capital cost per job at the Navan mine, which came on stream in the following year, was £93,800.[38] Direct employment in mining increased from 104 in 1964 to 2,100 in 1977 (about 0.2 per cent of the workforce).

Though small in extent, the type of employment provided does have attractive features. Mining employment is largely located outside the most prosperous regions, and therefore has a favourable impact on regional income distribution. The provision of employment in poorer areas can also have an important social impact, as it can stem the flow of local people to major Irish cities and abroad, preventing further social and economic decline. In certain cases, however, mine employment may only delay the process of decline. Such employment is finite, and the characteristics of regions suffering economic decline militate against alternative employment being available when a mine is exhausted.[39]

No reliable estimates exist regarding the extent of indirect employment created by the industry's purchases of goods and services. Kearns[40] claims that five additional jobs result from every one at the mine. This figure follows a generally accepted estimate,[41] but does not seem to reflect any detailed research into the question.[42] In fact a job multiplier of five seems grossly inflated. For example, Nickel et. al.[43] estimate that 2.0 is the job multiplier for the Canadian mining industry, while the Australian Mining Industry Council estimates the multiplier in Australia at 2.75.[44] The Australian and Canadian economies are in a position to supply a higher proportion of mining inputs than is the case in Ireland, and it is very unlikely that the job multiplier in Ireland would be higher than in those countries. Thus a multiplier of 2.0 would probably not overestimate the mining industry's indirect impact on employment. Total employment generatd by mining is therefore estimated at about 6,000 in 1977,or 0.6 per cent of the workforce.

Total wage and salary payments by the mining industry over the period 1964-77 were approximately £45 million.[45] It must be remembered that a large portion of these wages and salaries were expended in areas suffering from economic and social decline. They consequently possessed an added significance in that their expenditure probably rendered viable services (e.g. transport, retail outlets, schools) which would otherwise have been uneconomic. Thus a benefit accrued to the non-mining population, but the extent of any such 'spin-off' cannot be measured.

Creation of Additional Economic Activity through Linkage Development

Backward Linkage. Detailed information on the mining industry's local purchases of goods and services is only available for the period 1964 - June 1973; total expenditures during these years were £31 million.[46] Many of the inputs used in mining have been imported, particularly machinery and equipment. The Canadian origin of the companies concerned apparently influenced sourcing of equipment purchases, which have been made almost exclusively in Canada. However, it is extremely unlikely that the Irish mining industry could have provided a sufficiently large market to allow the economies of scale required for efficient domestic production of items such as underground mining equipment or concentrating machinery. The only real choice facing any investor (domestic or foreign) would have been whether to make the purchase in, for example, Britain, Canada or the United States. Assuming that Canadian suppliers were competitive, it is therefore unlikely that Ireland's interests suffered from the inclination of foreign investors to purchase goods in their home country.

The proportion of total capital expenditure being made in Ireland has risen sharply over the period since early mineral developments took place. For the most recent project completed, the Navan zinc/lead mine, the relevant figure was 56 per cent,[47] while the equivalent figure for the Tynagh mine was about 18 per cent.[48] Since almost all equipment purchases continued to be made abroad, the trend in other categories of expenditure has clearly been more dramatic than these figures suggest.

Two factors account for this trend. First, the companies involved in later projects have felt it necessary, for political reasons, to be seen to make as large a contribution as possible to the Irish economy. They have consequently followed a deliberate policy of using local sources of supply where possible. Second, Irish firms seem to have been slow to take advantage of the opportunities presented by mineral development. This lag in linkage development probably reflected the time required for local companies to acquire the relevant information, skills and capabilities. Indeed, information was apparently the key factor, as many of the opportunities available involved activities (e.g. shaft sinking, provision of structural steel) already being carried on by local firms. The proportion of current expenditures being made in Ireland has also increased over time, probably for the same reasons, and stands at about 80 per cent for currently operating mines,[49] as opposed to about 65 per cent during the early period of Tynagh's operations.[50]

It should be noted that the companies involved did not utilise commodities produced by affiliates. Had they done so, their attitude towards using domestic sources of goods and services might have been different.

Forward Linkage. All non-ferrous minerals produced in Ireland were exported in concentrate form, and consequently forward linkage was absent. The absence of the first stage of forward linkage, smelting and refining of concentrates, was not due to any reluctance on the part of foreign mining companies to undertake such activities. Since the mid 1960's, the Northgate group had made clear its commitment to constructing a zinc/lead smelter in Ireland. In 1968 Northgate established the Smelter Corporation of Ireland (SCI) to undertake relevant feasibility studies, and spent over £100,000 on this work during the next two years. By Spring 1969 SCI had bought a suitable site, and in March 1970 it applied for planning permission to build a smelter at Little Island, Co. Cork.[51] SCI did not in the event proceed with the project, for a number of reasons, including the following: (1) Northgate had apparently negotiated the involvement of a major European zinc smelting company in the project, as a source of capital and relevant expertise. This company withdrew at a late stage, threatening the project's viability. (2) The proposal to construct a smelter encountered serious opposition from local groups concerned with its environmental impact. (3) In 1970 a major zinc/lead deposit was discovered at Navan, Co. Meath, by Tara Exploration and Development Company Limited (see Chapter Three). This had an important bearing on the size and location of any zinc smelter constructed in Ireland. Northgate subsequently entered an agreement with Tara and a major Canadian zinc producer, Noranda Mines Limited, to study the feasibility of constructing an electrolytic zinc smelter in Ireland.[52] This arrangement was in turn superseded by the Irish government's 1977 decision to construct a smelter in association with New Jersey Zinc and the companies working the Navan deposit.[53]

## Net Foreign Exchange Inflow

Mineral exports as a percentage of total exports were at their highest in 1970 at 4.7 per cent, declining to 1.3 per cent in 1977, primarily as a result of a rapid rise in the value of total exports (almost 300 per cent between 1970 and 1976).[54]
        Table 2.3 calculates the net foreign exchange inflow resulting from Northgate's mining operations over the period 1966-72. Net smelter revenue represents the total value to Ireland of ore exported. Items representing expenditures in Ireland can be indentified, and thus the net foreign exchange inflow calculated. Precise data was available regarding all major categories of local expenditure: payroll costs, government royalties, local disbursements, and dividend payments to local shareholders. The only item which required estimation was the import content of domestic expenditure.[55]
        Table 2.3 indicates that net foreign exchange inflow amounted to 35.5 per cent of the value of export receipts over the period 1966-72. It was accounted for by government revenue (10 per cent), payroll costs (41 per cent), domestic purchases of goods and services (44 per cent), and dividend payments to local shareholders (5 per cent). Table 2.3 also show that the proportion of export receipts retained in Ireland increased substantially over this period. Two factors explain the change. Metal prices were depressed in 1971-2, and so a greater proportion of revenue was being directed towards current and capital expenditure (a substantial

Table 2.3 : Net Foreign Exchange Inflow, Tynagh and Gortrum Mines, 1966 – 72 (C$000s)

| | 1966* | 1967* | 1968* | 1969* | 1970 | 1971 | 1972 | Totals |
|---|---|---|---|---|---|---|---|---|
| Net Smelter Revenue | 10,245 | 15,442 | 16,118 | 15,986 | 19,646 | 13,187 | 16,715 | 107,339 |
| Local Disbursements | 2,203 | 2,268 | 3,344 | 2,981 | 6,210 | 7,469 | 5,689 | 30,164 |
| less Import Content of Local Disbursements** | 980 | 1,009 | 1,488 | 1,326 | 2,764 | 3,324 | 2,532 | 13,423 |
| Net Local Disbursements | 1,223 | 1,259 | 1,856 | 1,655 | 3,446 | 4,145 | 3,157 | 16,741 |
| Payroll Costs | 960 | 1,349 | 1,346 | 1,485 | 2,582 | 3,657 | 4,122 | 15,501 |
| Dividends to Local Share-holdings | - | - | 545 | 554 | 565 | 302 | - | 1,966 |
| Government Royalty | 420 | 738 | 738 | 887 | 834 | 61 | 263 | 3,941 |
| NET FOREIGN EXCHANGE INFLOW | 2,603 | 3,346 | 4,485 | 4,581 | 7,427 | 8,165 | 7,542 | 38,149 |
| - As % Net Smelter Revenue | 25.4 | 21.6 | 27.8 | 28.6 | 37.8 | 61.9 | 45.1 | 35.5 |

\* Only the Tynagh mine's operations are included in these years.
\*\* Import content of local disbursements was taken as 44.5%, an estimate derived from published company documents.

Source: Table 2.1; published company documents.

proportion of which was made in Ireland), and a smaller proportion towards dividends and undistributed profits (very little of which remained in Ireland). Second, as mentioned above, the proportion of capital and current expenditures being made in Ireland increased during these years.

The net foreign exchange effect disclosed in Table 2.3 would be augmented by any inflow of capital associated with Northgate's operations in Ireland. Over the period to 1965, Northgate expended some C$13 million in association with the Tynagh project. (After that date capital expenditure was financed from mineral sales.) Of this amount, some C$4.1 million was expended in Ireland.[56] Thus total net foreign exchange inflow was about C$42.3 million, or 34 per cent of total gross foreign exchange inflow.

Table 2.4 outlines the composition of foreign exchange outflows resulting from Northgate's operations. Almost 30 per cent of the outflow consisted of distributed and undistributed profits, indicating the significance of the low tax rates applied to mining companies in determining their balance of payments impact.

Provision of Scarce Resources through the Investment Process

Foreign mining companies played a crucial role in providing capital for mineral exploration and development, a point discussed above. They also played an important part in providing the technology and technical expertise required to apply that capital efficiently. The clearest example of the transfer of a particular technology to Ireland involved the application of large-scale trackless underground mining methods to low-grade orebodies. This technique had been applied and perfected in Canada, and was employed by Canadian-based companies to develop the Avoca, Silvermines and Tynagh underground mines.

The technology applied by foreign mining companies was apparently not inappropriate, as there is no indication that other than capital-intensive highly-mechanised mining methods could have been employed in working low-grade underground mines of the type established in Ireland.

Northgate's Tynagh mine offers an example of a situation in which availability of specific skills played a crucial role in permitting mineral development. Tynagh's ore was extremely complex. In

Table 2.4 : Composition of Foreign Exchange Outflow, Northgate Group, 1966 - 72

| Category | Amount (C$000s) | % of Total |
|---|---|---|
| Loan Repayments | 15,785 | 22.8 |
| Interest Payments | 2,829 | 4.1 |
| Dividends Paid Abroad | 7,866 | 11.4 |
| Undistributed Net Income* | 12,117 | 17.5 |
| Operating and Capital Expenditure Incurred Abroad** | 17,171 | 24.8 |
| Import Content of Local Expenditures | 13,423 | 19.3 |
| Total | 69,191 | 100.0 |

* This figure equals the sum of net profits plus depreciation and amortization allowances, less dividends, capital expenditure and loan repayments.

** This figure was derived by subtracting payroll costs and local disbursements from the sum of operating, exploration and capital expenditures.

Source: Tables 2.1 and 2.3; published company documents.

addition to the variety of rock types involved
(seven different categories were used to state ore
reserves in 1971), the ore minerals were contained
in a black/brown mud where sand-sized particles were
bonded by fine clay-like substances, making mineral
extraction very difficult. Five separate
concentrates, lead, zinc, copper, bulk lead/zinc and
'lead oxide' had to be produced; specialised
metallurgical skills were applied to permit
efficient flotation of individual ore minerals,
allowing profitable exploitation of a complex
deposit.[57]

Initially, much of the skilled labour utilised
in mineral development was supplied by the parent
companies, but most expatriates were quickly
replaced by Irish personnel. For example, when Avoca
Mines Limited commenced production in 1971, a high
proportion of skilled manual labour and nearly all
professional and managerial staff were Canadian;
seven years later, only a handful of Canadians
remained.[58]

Creation of External Economies

Little or no 'spin-off' effects resulted from the
transfers of skills and technologies described
above. Expertise in underground mining or in
metallurgy was of no relevance to other sectors of
the economy. Neither was it utilised by domestic
mining companies, which were exclusively involved in
quarrying, using open pit methods which had long
been applied.

The contribution of the mining industry to
infrastructural development has been modest as a
result of the already highly-developed state of
social and physical infrastructure in Ireland.
Nevertheless, the industry has had some impact. This
has in general been indirect, in that
infrastructural development has been carried out and
financed by state or local authorities, but has
depended for its viability on use by mining
companies. For the most part, such development has
also facilitated non-mining activity.

The channel in the port of Galway was deepened
in 1964 to allow export of Tynagh's ore, and the
port's tonnage limit was thereby raised from 2,000
to 5,000. Docks were enlarged and new storage
facilities constructed. Though the closure of Tynagh
will lead to a substantial decline in traffic,
Galway's port will continue in use, for example in
the importation of oil products. At Foynes, on the

Shannon estuary, a new jetty capable of taking ore ships up to 25,000 tons was constructed to serve the zinc/lead and barytes mines at Silvermines. Once again, this port is capable of handling non-ore cargo. In addition, its existence has enabled Foynes to take advantage of opportunities presented by oil exploration off Ireland's west coast, and six wells were drilled out of Foynes in 1978 alone. A new railway line was built to link Silvermines with Foynes, but it is unlikely that significant non-mine traffic utilises this section.

No information could be obtained regarding costs incurred or revenues received by public authorities in relation to provision of infrastructure.

## Creation of External Diseconomies

There is no indication that significant external diseconomies have been associated with mineral development in Ireland. Unemployment was high during the period under review, especially in the areas in which mining took place, and it is consequently unlikely that significant opportunity costs resulted from labour absorption. Little domestic capital was mobilised, but domestic investors were not denied the opportunity to become involved in the mining sector. As to whether minerals should have been conserved until Irish investors were able to grasp that opportunity, it is apparent that the obstacles to their doing so were deep-seated and unlikely to be quickly overcome (see above). Given that Ireland faced severe and immediate economic problems, the preference of policy-makers for immediate foreign-financed mineral development seems justified in economic terms; given the political imperatives facing Irish governments, it was probably inevitable. It is unlikely that mining activity exerted a major influence on domestic income distribution. Returns to national investors do not seem to have been affected, and it is unlikely that mining's demand for labour influenced wage levels significantly, given substantial unemployment. Miners' wages were above the national average (by 37 per cent in 1973), but the numbers involved were small (0.2 per cent of the workforce) and not very 'visible' within the national economy, and so the difference is very unlikely to have exercised any 'demonstration effect' on wage levels.

This analysis indicates that foreign-financed mining projects did contribute to the government's central aim of increasing employment and export receipts. Their contribution to employment was modest, reflecting the capital intensity of mining operations, the absence of mineral processing, and the lag in backward linkage development. There was apparently little opportunity to vary the labour intensity of mining operations, but the lag in linkage development could probably have been overcome had government adopted a more interventionist approach. It is clear that foreign mining companies would have been responsive to pressure to purchase goods and services locally, while government could have ensured that local suppliers were informed of the available opportunities.

Irish poicy-makers adopted the wrong criterion in assessing the contribution of foreign-financed mining projects to the balance of payments. The criterion applied was that of export receipts, rather than of net foreign exchange inflow, and consequently the implication of low tax rates for the balance of payments was not apparent. Had taxation rates been higher, so would the scale of net foreign exchange inflow.

The criterion of export receipts suited Irish politicians very well of course, for two reasons. First, it offered a more easily definable and readily understandable measure by which those politicians could demonstrate their 'success' to the Irish public. Second, it obscured the true implications, from Ireland's point of view, of the type of relationships which developed between them and the foreign investors concerned.

Notes

1. The official title of the twenty-six counties was changed in 1949 to the Republic of Ireland or Eire.
2. Dail Debates, 19 February 1936.
3. See, for example, the statements by government spokesmen, Dail Debates, 8, 9, February 1956.
4. Dail Debates, 8, 9 February 1956.
5. Interview with retired Geological Survey official, Dublin, 8 November 1978; Dail Debates, 25, 26 July 1962.
6. Dail Debates, 21 February, 25 July 1962.
7. When the Avoca project was initiated, the London Metal Exchange copper price stood at £430 per ton; on the day production commenced, it stood at £186 per ton.
8. This method involved extensive soil and water

sampling which provided sufficient information to
allow pin-pointing of likely targets for detailed
exploration.
9. Minister for Industry and Commerce, 'Report ...
for the six months ended June 30th, 1975, in
accordance with Section 77 of the Minerals
Development Act 1940' (Department of Industry
and Commerce, Dublin, 1975), p.9.
10. D.D. Derry, 'What Made Ireland Go', Paper
presented to the Prospectors and Developers
Association, 10 March 1969, pp. 5-6.
11. The 1959 amendment gave general applicability to
the 1956 concessions by extending the time limit for
commencement of production to 7 years.
12. Eighty per cent of the Northgate group's shares
were held by individuals or groups resident outside
Ireland, and nearly all of Northgate's expenditure
on exploration and development was financed from
abroad. Mogul of Ireland was 75 per cent owned by
its Canadian parent, while the remaining 25 per cent
was held by a company with substantial British
shareholding. The second Avoca development was
almost wholly Canadian owned, while both it and
Mogul were entirely financed by foreign funds. The
baryte project at Silvermines and the dolomite
development at Bennetsbridge were 100 per cent owned
and financed by US companies.
13. Local investors may not have been given the
chance to participate in projects financed by
wholly-owned subsidiaries of vertically-integrated
multinational corporations, such as Bennetsbridge,
but they were given opportunities for investment
elsewhere, and at least one company (Northgate) made
a deliberate attempt to mobilise domestic capital.
14. For a more detailed discussion of these factors,
see C.O'Faircheallaigh, 'The Role of Foreign
Investment in Mineral development: A Comparative
Analysis', unpublished PhD thesis, Australian
National University, 1980, pp. 61-3; J. Lee,
'Capital in the Irish Economy', in L.M. Cullen
(ed.), The Formation of the Irish Economy (Mercier
Press, Cork, 1969), pp. 53-63; J. Meenan, The Irish
Economy since 1922 (Liverpool University Press,
1970), p.63.
15. Some revenue would continue to accrue in the
form of royalties, but royalty rates were modest,
and royalties were only applicable to minerals under
state control.
16. Dail Debates, 11 April 1967.
17. Interview with the official concerned, Dublin,
22 November 1978.

18. Interview, Dublin, 8 November 1978.
19. M.O'Kelly, 'Legislation and the Irish Mining Industry', unpublished MBA dissertation, University College Dublin, 1978, p.89.
20. See, for example, Sinn Fein, The Great Irish Oil and Gas Robbery (Dublin, 1975), p.1, and Resources Study Group, Navan and Irish Mining: Documentation of a £850,000,000 Robbery (Dublin, 1971), Introduction, p.1.
21. G. Fitzgerald, Planning in Ireland (Institute of Public Administration, Dublin, 1968), p.15.
22. This acceptance was based on the attribution of the increased growth rates and declining emigration which characterised the years 1958-67 to the strategy of foreign-financed industrial development.
23. Special provisions relating to marginal mines were included in the Finance (Taxation of Profits of Certain Mines) Act, 1974. These provisions have already been applied in one case.
24. O'Kelly, 'Legislation and the Irish Mining Industry', pp. 139-49.
25. Dail Debates, 15 May 1962.
26. Dail Debates, 3 May 1967.
27. See, for example, Dail Debates, 3, 23 May 1967 and 29 June 1972.
28. See the statement by Mr Haughey, the then Minister for Finance, quoted above; interviews with officials of the Department of Industry and Commerce and of the Geological Survey Office, Dublin, 24 October, 8, 22 November 1978.
29. The twenty-year tax exemption was abolished by the Fine Gael/ Labour Party Coalition government in September 1973.
30. Indirect taxation has apparently been insignificant, as mining industry imports have been exempt from customs duties and domestic purchases exempt from sales tax.
31. Concentrates are sold to independent smelters at prices based on LME quotations.
32. This was established by examining information disclosed by the parent companies in their annual reports regarding their associated companies and principal investments.
33. Profits attributable to mining were C$8,358,000 or £3,248,000. (Canadian dollars are converted to pounds sterling on the basis of average exchange rates during the year in question). Royalties would, according to the mining lease, be £257,320. Derived from Minister for Industry and Commerce, 'Report', p.9; Table 2.1.
34. Derived from Table 2.1

35. Profits attributable to mining were C$8,216,000 or £3,186,000. Royalties would, according to the mining lease, be £251,740. Derived from Minister for Industry and Commerce, 'Report', p.9; Table 2.1.
36. Northgate Exploration Limited, Annual Report, 1975, p. 22.
37. Initial investment in The Tynagh mine was C$13,082,000, and subsequent capital expenditure on Tynagh and Gortrum was C$18,343,000 (Derived from Northgate Exploration Ltd.,Annual Reports, 1962-1977). To 1977, amortization and depreciation allowances totalled C$35,309,000 (See Table 2.1).
38. Total capital costs were £75 million, and 800 permanent jobs were created.
39. This may well be the case with Silvermines and Tynagh, for example.
40. K.C. Kearns,'Ireland's Mining Boom: Development and Impact', American Journal of Economics and Sociology, June 1976, p.263.
41. See, for example, J. McConnell, 'Mining - Its value to the National Economy', Development, October 1968, p.8; and H. Legge, 'Revival - Mining in Ireland' in H. Legge (ed.), Mining Ireland (Dublin, 1973), p.9.
42. Kearns gives as his source an article in a small provincial newspaper, the Wicklow People, while the other authors give no source for their estimates.
43. Nickel et al, Economic Impacts and Linkages of the Canadian Mining Industry (Centre for Resource Studies, Kingston, Ontario, 1978), p. xi.
44. Australian Mining Industry Council, Mining Review, November - December 1977, p. 2.
45. This figure is based on published company documents and on information provided by Avoca Mines Limited.
46. Published company documents.
47. Irish Press, 28 October 1976. This figure was confirmed by an industry source.
48. This figure is an estimate gained by calculating local expenditures during Tynagh's development period (1964-5) as a percentage of capital costs.
49. This is the proportion which applies to the Navan zinc/lead mine and the Avoca Mines copper mine.
50. Derived from Tables 2.1 and 2.3.
51. Northgate Exploration Limited, Annual Report, 1968, p. 22; Irish Times, 29 April 1970, 15 September 1971.
52. Northgate Exploration Limited, Annual Report, 1973, p. 17
53. Irish Press, 25 March 1977. New Jersey Zinc

subsequently withdrew from the project because of adverse conditions in the international zinc market. The smelter project is now in abeyance.

54. Central Bureau of Statistics, _Trade Statistics of Ireland_ (Government Printer, Dublin), various issues.

55. See the notes to Table 2.3.

56. Derived from Northgate Exploration Ltd., _Annual Reports_, 1961-1965; published company documents.

57. For references to literature describing the problems encountered at Tynagh and the solutions adopted, see C. Burton, 'Production and Reserves of Base Metals in Ireland', _Technology Ireland_, September 1977, p. 40.

58. Interview with Avoca Mines Ltd.'s Production Manager, Avoca, 9 November 1978.

Chapter Three

THE NAVAN ZINC/LEAD DEVELOPMENT: A CASE STUDY OF
FOREIGN INVESTOR - HOST COUNTRY RELATIONS

One of the many companies which undertook
exploration in Ireland in the wake of the Tynagh
discovery was Tara Exploration and Development Company Limited, a Toronto-based firm. Tara was closely
linked with Northgate, having been established by
the same small group of men. On 4 November 1970 Tara
announced that it had 'encountered mineralisation...in the first drill-hole of a new prospect near
Navan, Co. Meath.' It soon became apparent that
Tara's find was on a scale unprecedented in
Ireland's mining history. In early 1973 the Company
stated proven ore reserves at 77 million tons,
grading on average 11.0 per cent zinc and 2.7 per
cent lead. Tara had discovered what is now one of
the world's largest known deposits of zinc and lead.
    In the years after the Navan discovery, a major
political debate developed concerning the position
of the mining industry and its contribution to the
Irish economy. The scale of Tara's find generated
considerable public interest, and a number of
political groups took advantage of this situation to
demand changes in the structure of the industry. The
most influential of these was the Resources Study
Group (RSG), a Dublin-based organisation which
succeeded in greatly increasing public awareness of
the issues involved. Along with Sinn Fein (Gardiner
Street), a small socialist party, and with left-wing
members of the Labour Party, the RSG campaigned for
nationalisation without compensation of all foreign
mining companies. Other political groups, while
anxious to ensure continued foreign investment in
the industry, demanded abolition of the twenty year
tax exemption.

Ireland Changes its Government - and its Mineral
Policy

When the Fianna Fail party was defeated in the
general election held on 23 February 1973, it was
clear that the Fine Gael/Labour Party coalition
which replaced it would alter mineral policy in
important ways. Strong pressures in this direction
would certainly be generated by sections of the
Labour Party which, at its annual conference in
1972, had passed a resolution calling for the
nationalisation without compensation of the
country's mineral resources.[1]
    The newly-appointed Minister for Industry and
Commerce, Mr Justin Keating, indicated that he
regarded the formulation of a comprehensive policy
covering the development of Ireland's natural
resources as one of his most urgent tasks. However,
he ruled out the possibility of nationalising the
mining industry, for political rather than financial
or technical reasons:

> I don't advocate total public control of oil
> and gas and mines. I do not advocate this
> because I am a democrat. No Irish government
> has a mandate to do this. I am not going to ram
> down the throats of the Irish people what they
> do not want.

As Mr Keating pointed out to the Labour Party conference in 1973, whatever his personal preferences,
legislation would have to be framed in such a way as
to ensure the support of Fine Gael in both the
Cabinet and the Dail. For these reasons, it was
essential to find a middle road 'between the two
ends of the possible outcome, one end total nationalisation, the other end the previous policy of
total private control'.[2] Mr Keating's general
political beliefs also gave clear indication of the
policies he would pursue in relation to the mining
sector.[3]
    After a life-long involvement in the socialist
movement Mr Keating was deeply committed to its
ideals, but had strong views on how these should, or
should not, be implemented. Perhaps most
importantly, socialism had to be democratic. Thus
his determination not to 'ram down the throats of
the Irish people' measures which they did not want
and to press only for measures on which there was a
consensus within the Coalition. Keating's second
basic tenet was that efficiency should be the key-

note of any enterprise, public or private. High
standards of efficiency would have to be achieved by
the public and private sectors if Ireland's social
and economic problems were to be solved. He was
appalled, for example, by the pouring of British
Funds into projects such as Concorde and into firms
such as British Leyland. He was equally appalled by
the ignorance of and lack of feeling for the
business world which he perceived among the civil
servants of his Department (Industry and Commerce)
and described their lack of business skills as
'terrifying'.

The public sectors of countries such as Britain
and Ireland left a great deal to be desired, therefore, in terms of efficiency. On the other hand,
Keating believed that private enterprise and the
capitalist system within which it operated had
failed to solve the urgent problems facing Ireland
and the world, for example, poverty, environmental
despoliation, and energy shortages. Thus there was a
great need for effective planning within a
democratic framework. This necessitated the active
involvement of the state in key areas on the
economy.

Keating saw another, and complementary, reason
for direct state participation in business. When the
'oil crisis' occurred in October 1973, the Irish
government set out to find out where Ireland's oil
was coming from, and found that it could not do so.
Oil was being traded among the multinationals on the
high seas, and the original source was impossible to
discover. In addition, the crisis called attention
to the pricing policies of multinational oil
companies in Ireland, policies Keating believed
might be aimed at avoidance of Irish corporate
taxes. In general, he considered that private
enterprise 'may have very strong reasons...for not
being completely frank with the state'. Direct state
involvement in commercial activity was necessary if
government was to overcome its ignorance in such
areas.

In addition, competition between private and
public enterprises would raise the general level of
efficiency in the economy; it would force the public
sector 'to apply the efficiency, the standards, the
methods, even the accounting of private industry to
the public sector'.[4] It would end the woeful
ignorance of public servants concerning the business
world, allowing them to monitor private enterprise
effectively and to ensure that its activities
furthered government planning aimed at dealing with

social and economic problems.

What Keating wanted was a system within which the public and private sectors interacted constantly, adding to the efficiency of the former and the social utility of the latter. The implications for the mining industry were clear: Mr Keating would aim to actively involve the Irish government in the industry, presumably through equity participation in existing companies and eventually through direct initiatives by the state in mineral exploration and development. In general, it was obvious that Keating would have very little sympathy with the laissez-faire policies of his predecessors.

Throughout the summer of 1974 speculation continued as to precisely what initiatives Keating would take. It was reported that an interdepartmental committee, set up by the Fianna Fail government in 1971 to review mining taxation and royalties, particularly the twenty year tax concession, had been ordered to extend its terms of reference. Legislation recently implemented by other governments was to be examined, and a policy formulated which would take account both of Ireland's special needs and of current international trends.[5] When this committee reported, it did not recommend abolition of the twenty year exemption. A substantial minority favoured abolition, but Department of Finance representatives who dominated the committee held the opposite view, and their viewpoint prevailed. Keating rejected the committee's recommendation, as he believed that Ireland's bargaining position had improved to such a degree that it was no longer necessary to offer generous tax concessions and because he felt that international developments had left Ireland very much out of line with other mineral producers in terms of tax concessions offered.[6]

Keating undertook the 'wheeling and dealing' necessary to gain Cabinet support for his initiative, but encountered considerable opposition. Mr Tully, the Minister for Local Government, who had long been a champion of mining interests, opposed the measure throughout, as did a number of other ministers. Keating nevertheless succeeded in obtaining Cabinet ratification, though the final decision was far from unanimous.[7]

The government announced its decision to end the twenty year tax concession on 25 September 1973, before Keating had outlined any general policy on mineral development. It stated the view that Ireland's mining industry, 'having reached its

present state of development should now make a
suitable and equitable contribution towards the
ever-increasing cost of promoting national development'. This contribution would be obtained through
the imposition of 'normal taxation' (i.e. 50 per
cent of profits) 'after deducting certain allowances
in respect of expenditure under such headings as
exploration, prospecting, development and new plant
and machinery'. It also stated that negotiations
would take place with the mining companies to fix a
sliding scale of royalties.[8]

The government's decision was met by a torrent
of criticism by mining industry spokesmen, who
claimed the abolition of the exemption would deter
foreign investment in mineral exploration.[9] A study
which examined exploration activity in the wake of
the decision has indicated that these fears were
unfounded. This study (by M. O'Kelly) shows that the
exploration industry, dominated until about 1970 by
small companies and by independent 'wildcatters',
thereafter attracted a higher proportion of large
multinational mining concerns. From 1971 to 1977,
the number of such companies on the Irish scene
grew without interruption. O'Kelly concluded that
Keating's initiative 'was a non-event as far as they
were concerned'.[10] In his view, stability of
operating conditions and more concrete evidence of
major mineral occurrences were of greater importance
to such companies than were tax incentives. The
Navan discovery provided the concrete evidence, and
it is clear that the ending of the tax exemption
was not seen as indicating instability in operating
conditions in Ireland.

The Finance (Taxation of Profits of Certain
Mines) Bill, 1974, circulated on 1 April of that
year, officially terminated the mining companies'
exemption from income and corporation taxes. It
stated, however, 'that relief [would] be provided
for marginal mines which might otherwise cease to
operate because of the potential tax burden'. Much
of the rest of the Bill was devoted to outlining the
allowances which the mining companies would be permitted to offset against taxation. Section 2 provided that all expenditure on exploration would
qualify for immediate allowance, as would development costs for any mine brought into production.
Section 6 provided for an additional special
allowance of 20 per cent of all expenditure on
exploration. Section 7 provided that 120 per cent of
the value of all new plant and machinery could be
immediately written off against taxation, while

Section 8 enabled expenditure incurred through the purchase of mineral deposits to be offset against taxation.[11]

These provisions have two principal effects. First, they exempt from taxation profits equal to 20 per cent of a mining company's expenditure on exploration, plant and machinery. Second, they postpone tax liability for a period determined by the scale of capital expenditures and the level of profits; in this instance the loss to government is represented by the notional interest on deferred revenue. It is not possible to predict precisely what the impact of the 1974 Act will be in particular cases, but its provisions will certainly soften the effect of the withdrawal of the tax exemption. In addition, it is evident that these provisions are being interpreted generously by the taxing authority, the Revenue Commissioners (a division of the Department of Finance), and that this will significantly increase their impact. As mentioned in the previous Chapter, an element of double depreciation was involved in Northgate Exploration Limited's application of depreciation and amortization allowances over the period 1962-77. Closer examination of its accounts showed that this practice apparently did not occur prior to 1973. In that year, the sum of allowable expenditures was C$29,312,000, substantially more than that of depreciation and amortization allowances (C$22,785,000). Four years later, the sum of allowable expenditure was C$31,425,000, while that of depreciation and amortization allowances had jumped to C$35,309,000.[12] Over the period 1973-7, the provisions of the 1974 Act were apparently applied by the Revenue Commissioners in a manner calculated to afford Northgate a significant additional measure of tax relief.

Thus the tax liability of mining companies has not increased as substantially as would first appear. This situation apparently results from decisions taken by civil servants in the Department of Finance rather than by politicians. According to Mr Keating, neither he nor his Department had any input into the precise form of the new legislation; he was able to influence its broad character, but the 'technicalities' of taxation provisions were regarded as purely a matter for the Department of Finance. He has also stated he knew the Revenue Commissioners, with the agreement of the Department of Finance, would draft a mild piece of legislation and interpret its provisions 'very generously'.[13]

The Navan Zinc/Lead Development

What explains the behaviour of the civil servants concerned, and Keating's failure to seek more stringent provisions?

As mentioned earlier, the Department of Finance opposed the termination of the twenty year tax exemption, and its officials apparently decided they would at least partly counteract its withdrawal. What motivated them to act in this way? Two possibilities suggest themselves. First, these individuals may have continued to hold attitudes towards foreign mining companies of the type discussed in the previous Chapter, and consequently believed it was still necessary to grant tax concessions if mineral exploration and production was to be maintained. Second, Department of Finance officials may have resented Keating's intrusion into a policy area they regarded as their own preserve, and been determined to assert their authority. Both factors may well have been involved.

Keating's failure to act was apparently due to his belief that he was fortunate to have secured implementation of any taxation measure, given the depth of opposition within the Cabinet, and that attempts to obtain more stringent provisions would be fruitless.[14] Thus the ability of civil servants to act in the manner described above was partly due to the continued adherence of some politicians to the attitudes described earlier. It also reflected the fact that their actions were not, and are not, public knowledge in Ireland. The effect of a twenty year tax exemption is easily grasped by the layman, but that of investment and depreciation allowances and of the manner in which they are interpreted is not. In addition, relevant information is difficult to come by, as the Revenue Commissioners do not publish details of their transactions with individual companies, and since the annual reports of Canadian mining companies are not freely available in Ireland.[15]

One question remains. Given these comments, why was the twenty year exemption ended? It seems clear that the political imperative was important; the pressure generated by the Resources Study Group, Sinn Fein, and the left wing of the Labour Party paid off, though the inability of these groups to influence decision-makers in the Department of Finance rendered their victory rather a hollow one. From Keating's point of view, the main importance of the initiative seems to have been tactical: it informed the mining industry, in the most dramatic fashion possible, that he meant business and that changes of a more significant nature were on the way.

## Bula Limited, and Irish Mining Legislation

Before proceeding further, it is necessary to explain developments in the area of mineral ownership in the wake of the Navan discovery. As pointed out above, the Mines and Mineral Development Act 1940, had provided a workable system of mineral ownership and was an essential element underlying mineral developments in the 1960s. Section 14 of the Act provided that

> Whenever it appears to the Minister [for Industry and Commerce] that there are minerals...[which] are not being worked or are not being worked efficiently...the Minister, with the consent of the Minister for Finance, may by order...compulsorily acquire such minerals or compulsorily acquire an exclusive mining right in respect of such minerals.

Section 22 allowed the Minister to grant to any person a licence to work minerals thus acquired. In effect, these sections ensured that if and when an exploration company discovered minerals, the Minister could acquire those minerals and permit that company to work them. He was not required to do so, but by the late 1950s it was assumed that he would. This assumption underlay the whole exploration effort from 1959 onwards, as it ensured that claim-jumping would be impossible and that successful companies would enjoy their just rewards.

In December 1970, Tara Mines Limited applied for a mining lease for that portion of the Navan minerals which were in state hands. But part of the deposit was privately owned, and Tara commenced negotiations with a local farmer, Mr Wright, in order to purchase the land involved. Tara offered Wright the agricultural value of the land, around £80,000; the company felt that it was the only bidder in the field, and therefore need go no higher. It was wrong. Another company, Bula Limited, made Wright a much more generous offer, which he accepted. On 19 March 1971 ownership of the land and mineral resources were vested in Bula.

On 22 March 1971 it was revealed that the Minister for Industry and Commerce, Mr Lalor, had signed a compulsory acquisition order under the 1940 Act taking the minerals concerned into state owner-

ship. This would defeat Bula's purpose in buying the
land as Lalor could simply turn the minerals over to
Tara. Bula Limited decided to contest the Ministerial order in the Courts, claiming that it was
invalid on technical grounds and, more generally,
that Section 14 of the 1940 Act, which allowed for
compulsory acquisition of minerals, was unconstit-
utional in that it infringed rights of private property.[16] The implications of this latter claim were
far-reaching. If upheld, it would invalidate all
orders (18 altogether) previously made under the
Act, and undermine the mineral exploration industry
in Ireland.

Bula's case commenced in the High Court in July
1972. But before then, another important development
had occurred. By March 1971, disturbed by the Bula
action and anxious to develop its deposit, Tara was
pressing the Department of Industry and Commerce to
issue a mining lease. Department officials were
loath to do this before the ownership issue was
settled, wary of any other legal pitfalls which
might await them.[17] Tara made representations to
Fianna Fail politicians with whom it had contacts,
and the Minister for Industry and Commerce gave Tara
a written undertaking that it would be granted a
mining lease. This document was (perhaps appropriately as things turned out) dated 1 April 1971; it
did not specify the terms on which the lease would
be granted.[18] Armed with this commitment, Tara began
to develop that part of the orebody not in dispute.

Bula Limited won its High Court case in April
1973, and the appeal brought against this decision
in the Supreme Court by the Minister for Industry
and Commerce, in March 1974. The High Court ruling
declared the Ministerial order invalid on the proc-
edural grounds only, and ruled that the 1940 act was
constitutional.[19] However, shortly after coming to
office Mr Keating obtained legal advice which
indicated that the constitutionality of the Act
would depend on the original owner being compensated
to the full commercial value of the ore in the
ground.[20] Thus in the Navan case, for example, the
state would have to pay farmer Wright the estimated
value of the nine million tons of lead/zinc ore
which lay under his land before it could permit Tara
to mine it. This advice was to impose an important
constraint on government action, the significance of
which will later become apparent.

Nothing had yet been revealed concerning government policy on the crucial questions of state
equity participation in mining ventures and the

level of royalties to be imposed; both would be
negotiated with the companies concerned. Discussions
took place with Tara Mines Limited, which was
anxious to finalise arrangements for the granting of
a mining lease, as early as October 1973. However,
the Department of Industry and Commerce was unwilling to issue a lease until litigation over ownership
of the Navan minerals had ended. In addition, the
government needed more time to prepare for negotiations with Tara.

Most importantly, Keating wished to enhance the
expertise available in his Department, Industry and
Commerce. He persuaded the then Secretary of the
Department to retire early. In his place he
appointed Mr Joe Holloway, formerly trade secretary
with the Irish embassy in London and a person with
considerable knowledge in the area of mineral
resource policy. Keating strengthened the powers of
the Secretary, and improved Holloway's access to
himself. He also doubled the number of relevant
technical personnel in the Geological Survey Office,
and arranged to draw on the expertise available
through the government-run Institute for Industrial
Research and Standards.

Keating would have liked to do more, particularly in hiring specialists from academic life and
private enterprise on a short-term basis, but the
Department of the Public Service vetoed this possibility. Keating had earned the ire of this Department by refusing to implement the recommendations
for civil service 'reform' contained in the Devlin
Report. In Keating's view, this report was attempting to achieve an even more 'closed' civil service
structure, whereas he believed that greater openess
and interaction between the service and the academic
and business worlds was required. Keating was also
unpopular with the civil service establishment
because he had raised Holloway to the position of
Secretary above the heads of senior career civil
servants within the Department of Industry and
Commerce. With his emphasis on technical expertise
and external recruitment, Keating was seen as a
threat by the administration-oriented career civil
servants who dominate the Irish Civil Service, and
on that rock many of his proposals foundered.[21] Thus
the Minister's attempts to increase Ireland's progress along the relevant 'learning curve' were ham-
pered by factors which were quite independent of any
relationship between host country authorities and
the foreign investors concerned.

Other preparations also took time. In November

1973 the Coalition did not have an agreed mineral resource policy, and formulation of a policy position was essential before negotiations could commence. The interdepartmental committee which had reviewed mining taxation was then examining the wider area of general mineral policy, and was assessing measures adopted by other governments. This committee was due to report in May 1974. Its recommendations would then be submitted to the government, which would have to decide whether to accept, amend or reject them.

While these preparations were underway, a major battle developed for control of Tara Exploration and Development Company Limited. Since its Navan find, which had raised Tara to the status of a large mining concern by international standards, a takeover bid for the company had been expected, and in February 1974 it materialised. Cominco Limited, which holds a duopoly position in the Canadian zinc industry with Noranda Limited, and Charter Consolidated, the London arm of the Anglo American Corporation of South Africa Limited, made a bid which valued Tara at C$162 million. Tara's management, determined to retain control of the Company, rejected the offer. In so doing, Hughes, McCarthy and Gilroy, whose personal, family and trust interests accounted for 20 per cent of Tara's shares[22] turned down a tax-free capital gain which might have amounted to as much as C$25 million[23]. They fought a tenacious battle against the take-over bid, but the need to issue shares to raise initial development capital had seriously weakened their hold on Tara, and they were forced to seek the assistance of Noranda Limited. Noranda responded and the bid was fought off, but in return for its help, Noranda, already a minority shareholder in Tara, received an option to purchase a controlling interest in the company at a later date, an option it exercised in March 1978.[24]

From Ireland's point of view, what was most disturbing about these events was the total incapacity of any Irish agency to influence the outcome of a battle for control of the country's most valuable single natural resource. On 18 February, the government stated publicly that there was no action it could take, as the company involved was under Canadian jurisdiction. In consequence Keating, who was anxious to intervene and prevent the take-over, was unable to do so.[25]

The Navan Zinc/Lead Development

The Bula Agreement

By July 1974, negotiations were underway between the Department of Industry and Commerce and Bula Limited concerning the development of Bula's section of the Navan orebody. Keating was well disposed towards Bula, a wholly Irish-owned company, and the Bula directors, having pulled off a coup of impressive magnitude, were doubtless prepared to accommodate the Minister's wishes. At the end of July agreement had been reached between the two parties. The text of the agreement, entitled Bula Limited – Inter-Party Agreement, was later released by the Department of Industry and Commerce.

The financial provisions were as follows. Corporation tax would apply at the standard rate (50 per cent), but as Bula's minerals were in private ownership no royalties would be charged. Bula Limited would hand over to the state, free of charge, 25 per cent of its equity (Article 2.02). In addition, the company would sell to the Minister a further 24 per cent of its shares, at a price fixed by independent consultants (Articles 2.01, 3.01(a)). The purchased shares would not, however, have voting rights (Article 9.01(1)). The Minister would appoint two out of the company's seven directors (Article 4.05). Article 8.04 provided that Bula Limited would 'distribute by way of dividend the entire profits of such financial year as shown in the audited accounts of the Company remaining after taxation applicable thereto'. This requirement would not apply where limits on dividend payments were contained in loan agreements or where its auditors advised the company that its commitments made dividend payments inadvisable.

Article 8.05 provided that if and when a zinc refinery was established in Ireland, Bula would supply concentrates to it 'on normal and reasonable commercial terms'. Bula was guaranteed an equity stake in any smelting venture, but would not be compelled to take up this equity (Article 9.07(a)). Considerable emphasis was placed on ensuring Ministerial access to corporate information; for instance, Articles 6.01 and 8.07 called for the Company to supply full information on its business activities, and to give the Minister access to its 'properties, books and records'.

In theory, the state 'take' under these arrangements would amount to almost 75 per cent of profits. However, 12 percentage points of this would be accounted for by dividends from shares purchased

at their full commercial value. In addition, the
provisions of the Finance (Taxation of Profits of
Certain Mines) Act 1974 would ensure that the 'free'
state 'take' of 62.5 per cent would in practice be
significantly below that figure.

On the other hand, Keating had ensured
inclusion of provisions designed to further his
overall strategy, outlined above. It will be remembered that Keating's major priorities were to ensure
direct state involvement in productive enterprises,
to educate civil servants in the skills practiced by
private enterprise, and to ensure effective government monitoring of private businesses. A state
shareholding of 49 per cent in Bula represented a
significant step towards the first objective, though
a far cry, of course, from the ultimate aim of a
state-owned independent mining company. The nomin-
ation to Bula's Board of Directors of two civil
servants was a step towards development of the
required skills in the civil service. Their presence
on the Board, and the access to information guaranteed under the Agreement, would ensure that Bula's
activities could be closely monitored. From this
point of view, the Bula agreement was a significant
triumph for Keating.

Following the conclusion of terms with Bula,
and given that Tara had long been impatient to
commence mining at Navan, it was expected that the
Minister's negotiations with the company would
rapidly reach a satisfactory conclusion. In fact,
the two parties were to fight a protracted battle
before arrangements for the granting of a mining
lease were made.

## The Negotiation of the Tara Agreement

By 8 July 1974, serious negotiations were under way
between the government and Tara. On that day, Mr
Keating outlined his proposals to the company's
representatives. These were similar to the terms
accepted by Bula, with an important exception. Since
the bulk of Tara's minerals were state-owned, the
Minister proposed to impose a royalty of 10 per
cent. The total state 'take' under such an arrangement would be 77.05 per cent. Mr Keating also sought
non-financial terms of the type agreed with Bula,
providing for disclosure of information, appointment
of government-nominated directors, and the diversion
of concentrates to an Irish smelter.

At this stage, Tara's position was very far
from the Minister's. The single most important issue

from its point of view appears to have been that of state participation. The company was apparently prepared to accept a state 'take' in the region of 60 per cent of profits, but it is evident that it was opposed in principle to equity participation by the Minister.[26] To understand the company's stand on the issue of state participation, it is necessary to keep in mind the character of its senior management, which remained unchanged in composition after the Noranda take-over. Hughes and his associates were unusual in the mining industry in that they both managed Tara and were also heavily involved in ownership of the company. This circumstance gave Tara's senior management an involvement in the company's affairs which was unusually personal and deep. They can hardly have been pleased at the changes in taxation policy initiated by Keating, and now, not content with taking profits, Keating wanted to control part of the company, appoint directors to the Board, and dig out information from company files and records. To individualistic entrepreneurs who had fought hard and denied themselves a personal fortune to prevent a private competitor from gaining control of Tara, the prospect of state influence on such a scale must have been, to put it mildly, most unwelcome.

Tara's management apparently did not realise that Keating's commitment to state participation was as strong as their own opposition to it. As mentioned above, the concept of state involvement in the private sector was an integral part of Keating's whole political philosophy; and there was another, more immediate reason why he would fight hard to obtain state participation in Tara. In 1971, a significant discovery of natural gas was made off Ireland's south coast. This had spurred interest in the country's offshore resources, and by 1974 a large number of multinational oil companies were anxious to commence exploration. Keating had promised to release the terms under which exploration would be permitted by early 1975. Knowing that the oil companies would be watching his negotations with Tara closely, he was determined to be seen as a tough bargainer. The government planned to demand a free equity holding of 25 per cent in any commercial oil finds made, and Keating felt that if Tara did not accept such a condition the oil majors certainly would not.[27]

Given these attitudes and perceptions, a head-on encounter between the Irish government and Tara was almost inevitable.

# The Navan Zinc/Lead Development

After 8 July, a total of seven meetings was held between the company and officials of the Department of Industry and Commerce, without result. On 16 August, the government and Tara clashed in dramatic fashion. The company announced it was closing down operations at Navan, and that it had issued a writ against the Minister of Industry and Commerce, suing for a mining lease and for damages for losses caused by the delay in granting it. The writ issued by Tara asked for a court ruling ordering the Minister to grant a licence on 'reasonable terms', which it claimed it was entitled to as a result of the undertaking given by Mr Lalor, the then Minister for Industry and Commerce, in April 1971. It also asked for a ruling that the Minister would be going beyond his powers in demanding state participation as a condition of granting a mining lease.[28]

Tara claims that Keating put forward his position on 8 July as non-negotiable, that it therefore felt that further discussions were futile, and that drastic action had to be taken to protect its shareholders' interests.[29] Tara apparently took Keating at his word, but why? Keating admits that he put his demands forward as 'final' on 8 July, but claims that all negotiators, whether in the boardroom or the market-place, present their initial offer as 'final'.[30] It does seem strange that Tara did not consider the possibility that he was engaging in a bargaining ploy. State negotiators almost inevitably start off the bargaining process by adopting an extreme position, presenting this as their 'final offer' in order to seem intimidating.[31]

Persuaded of the Minister's intransigence, Tara's management adopted an aggressive bargaining strategy, closing the mine and suing the Minister, actions which Tara's officials regarded as perfectly proper to protect, preserve, and promote the legitimate interests represented by the Company.[32] This strategy was likely to put considerable pressure on Keating to compromise for two reasons. First, pressure for a settlement would be generated by the laying off of 500 workers. Second, Tara's writ called for an unspecified sum in compensation for the delay in granting a mining lease, a sum which Keating was advised might reach £30 million.[33] The loss of public moneys to the tune of £2.5 million as a result of the Avoca failure had created such a reaction that it would be almost twenty years before an Irish Minister again contemplated state involvement in a mining concern. If Keating lost £30

million the Fine Gael Labour Party Coalition might
be destroyed as a political force for an equally
long period. Some of Keating's colleagues were certainly likely to take this view and, realising how
much they had to lose, put pressure on him to compromise. A number of Ministers and also certain Fine
Gael backbenchers were uneasy at Keating's treatment
of the mining companies. The Coalition was under
strain as a result of this and other issues, and
Tara may have felt the pressure exerted by this
strategy would break the government before it broke
the company itself. The company's officials may also
have believed that restraints imposed on government
action by the nature of the Irish Constitution
would prevent any retaliation against it, and consequently that Tara would emerge victorious from this
conflict. The nature of those restraints is examined
in detail below.

Keating felt that Tara's strategy reflected
influence exerted in Tara's boardroom by Noranda. He
believed that many Canadian mining companies,
including Noranda, were opposed to the idea of state
involvement, regarding it as inimical to their corporate interests, and did not want government
representatives on their boards or guaranteed state
access to corporate information. As a major shareholder in Tara, Noranda might be expected, Keating
believed, to use its board representation to ensure
that Tara adopted a similar approach to the issue of
state participation. Keating feels that this led
Tara, which should have realised that he couldn't
budge on the issue of state participation, to pursue
an irrational strategy, thereby doing itself
considerable harm in the longer term. [34] This belief
may have led the Minister to adopt a more entrenched
position after the mine closure occurred.

Two points must be made here. First, whether
Keating's assessment of Noranda's position was
correct or not, it seems certain that Tara's management needed no encouragement to resist state participation. If its strategy was irrational, then this
reflected the attitudes of the company's own staff.
The second point is that, as mentioned above, Tara
apparently did not realise the importance of the
participation issue to Keating. Obviously, a neg-
otiator should never take it for granted that his
opponent sees his point of view.

On Monday 19 August, Tara Mines Limited
instructed contractors on its Navan site to wind up
their operations. The government took the view that
Tara was exerting pressure on Mr Keating, and gave

the Minister its full support.[35] On 30 August Keating delivered a forceful speech which was interpreted by the press as threatening Tara with nationalisation if it did not come to an agreement. [36] Given what was said above regarding Keating's attitude to nationalisation and regarding constitutional protection of private property rights, it is unlikely that this option was seriously considered. An alternative course of action might have been to offer a mining lease to Tara on the government's terms (thus fulfilling the obligation arising from the undertaking given on 1 April 1971), and, if the company rejected those terms, to offer the lease to some other mining concern. A host of difficulties would have arisen as a result, however. First, Keating's legal advice was that the obligation was to offer a lease on 'reasonable' terms, and clearly Tara could have initiated endless litigation as to what constituted, or did not constitute, 'reasonable'. Second, Tara could have sued the Minister for compensation in lieu of its investment in the Navan mine, thereby placing him in an extremely awkward situation politically. (Once again the spectre of Avoca raised its head!) Third, the Minister would have had to compulsorily acquire that portion of the Navan deposit privately held by Tara. As noted earlier, Keating had been advised that any such action might require payment of very substantial sums in compensation.

Thus Keating was enmeshed in the legal and judicial constraints of the Irish constitutional system, and he in fact took the only course open to him - he sat back and did nothing.

By December, the Minister had countered Tara's legal moves by himself asking for a High Court ruling on certain questions of law pertaining to Tara's writ. Keating's principal motive in initiating this litigation seems to have been to defer the hearing of Tara's case; as will become apparent, he had come to believe that the longer the Navan mine stayed closed, the better his bargaining position would become. Clearly, the Irish legal system could work for the Minister as well as against him.

By this stage, Tara's position was weakening. In earlier years, assured of the Fianna Fail government's co-operation, Tara had spent freely on the Navan development. But it had not anticipated significant delays in bringing its mine into production, and by late 1974 the company was deep in debt and going deeper with every month that passed.

It therefore needed to reach an agreement very badly indeed. Apparently Tara's management made a fundamental miscalculation if it believed that the Coalition government would break before the company did. Keating claims that in late January 1975 Tara came to him and agreed to compromise on the issue of state participation.[37] Tara's viewpoint would probably be different, and its officials would doubtless point out that Keating made important departures from his earlier position. This was in fact the case, and reflected the Minister's own weaknesses.

One of these was the lack of support for his position among members of his own Labour Party in Cabinet, a lack of support which, with the exception of one individual, had nothing to do with the policy issues involved, but rather related to personal and broader political relations between Mr Keating and his colleagues.[38] In addition, Keating believed that the sympathy of one Cabinet member lay with Tara, and that this individual was keeping Tara informed of all relevant discussions and decisions in Cabinet. This represented a tremendous weakness from Keating's point of view, as he felt there could be no secrecy from Tara regarding any matter which had to go before Cabinet. Keating would not, for example, have his minimum acceptable position endorsed by Cabinet, as he believed Tara would learn of it, placing the company in a very strong bargaining position.[39]

As we have seen, another major constraint facing Keating arose from the nature of the Irish Constitution and the weaknesses inherent in the 1940 Minerals Development Act. In effect, Keating could not pose nationalisation even as an ultimate sanction, a situation different from that of most host country governments. Neither could he take compulsory acquisition measures, because of the political restraint represented by the need to pay very large sums in compensation. In combination, these constraints meant that if Tara did not work the Navan mine, nobody would. And while Keating might be able to achieve delays, he could not hold up a development on the scale of that mine indefinitely.

Both parties were therefore in weakening positions, and it was likely that a compromise would be reached. A few days before the Supreme Court was to give its ruling on the Minister's writ, it was announced that an agreement had been concluded between Tara and the Irish government.

The Navan Zinc/Lead Development

The Tara Agreement

The terms of the arrangement, contained in two documents, an **Agreement** and a **Mining Lease**, were as
follows. The state would receive, free of charge, 25
per cent of the shares of Tara Mines Limited, the
operating company, but there would be no purchase of
shares as there had been in the case of Bula
Limited. Royalties would be charged on pre-tax
profits at a rate of 4.5 per cent (**Mining Lease**,
p.16), and profits would be subject to corporation
tax (then 50 per cent) after deduction of royalty
payments, exploration expenditure, and allowances
permitted under the Finance (Taxation of Profits of
Certain Mines) Act 1974. Tara would turn over to the
state, free of charge, that portion of the Navan
minerals (about 6 million tons ore) in its
possession (**Agreement**, B). As in the Bula agreement,
a provision regarding the payment of dividends was
included, though its terms were less stringent: not
less than 50 per cent of the accumulated balance on
the Profit and Loss Account would be distributed in
each year. It was left to the discretion of the
Minister, rather than to the company's auditors (as
in Bula's case) to determine circumstances in which
it would not be advisable to so distribute profits
(**Mining Lease**, p.17).
   The Minister would appoint two directors to the
Tara Mines Limited Board, or 25 per cent of its
members, whichever is greater (**Agreement**, 3.03). As
in the Bula agreement, clauses were inserted guaran-
teeing Ministerial access to Tara Mines Limited's
'property, books, and records', and also requiring
the company to provide such 'additional financial
and operating data and other information as to the
business and properties of the Company as the
Minister may from time to time request' (**Agreement**,
6.06; **Mining Lease**, p.5). These requirements apply
to the operating company only, and the Minister will
have no guarantee of access to information relating
to or held by the parent company, Tara Exploration
and Development Company Limited. As in Bula's case,
Tara agreed to supply concentrates to an Irish zinc
refinery 'on reasonable terms', if requested to do
so. Tara would be offered an equity holding in any
such refinery, but would not be compelled to take it
up.
   The overall state 'take' under these arrangements will stand at a nominal rate of 64 per cent.
Because of the operation of the 1974 Act, the real

level of state revenue will be significantly lower.
     Tara's senior management did not accept the new
situation with good grace. By March 1978, it was
believed that Hughes, McCarthy, and Gilroy had
almost completely liquidated their personal shareholdings in Tara. Hughes resigned from the Tara
Mines Limited Board before the government-appointed
directors took up their positions.[40]

## The Conduct and Outcome of Negotiations

Two interlocked but separate issues were at stake in
the negotiations between Tara and the Irish authorities. The first involved the percentage of Tara's
profits which would accrue to the Irish government,
the second related to the degree to which that
government would participate in, and exercise
influence or control over, Tara's corporate
decision-making process. These issues were interrelated for two reasons. First, the Irish government
saw equity participation in the operating company as
a means of exercising influence and/or control, but
such participation would of course increase the
proportion of Tara's profits accruing as government
revenue. Second, Keating believed that the Irish
Exchequer would suffer financial loss in the longer
term if a measure of control were not obtained.
     Mr Keating achieved his principal objective by
obtaining a 25 per cent equity participation in Tara
Mines Limited. In addition, he gained acceptance of
government-appointed directors on the company's
Board, and of guaranteed Ministerial access to Tara
Mines Limited's records and files. He gained no such
privileges in relation to the parent company, but
the practical significance of the distinction
appears to be limited. The Minister might be
prevented from obtaining relevant corporate
information, but as all payments of profits, royalties and dividends will be assessed on the income of
Tara Mines Limited, and as significant transfers of
goods and services do not occur between Tara Mines
Limited and its parent,[41] this limitation is
unlikely to have important financial implications
for Ireland.
     The Tara agreement also gave the Irish govern-
ment considerable influence over the distribution
and use of corporate funds, as a result of the
provisions which demand payment of dividends at a
certain level and allow the Minister discretion in
determining when dividends should not be paid. In
this regard, the position of Tara Mines Limited will

be quite different from that of Northgate Exploration Limited's Irish subsidiary, IBM, whose use of
funds was entirely at the discretion of Northgate's
Board of Directors, located in Toronto. Tara's
parent company will of course be free of government
influence in this area. In other words, while the
investing company remains free to use the funds it
receives as it wishes, the Minister can play a
significant role in determining the extent of these
funds at any time.

The Irish government made important concessions
in the area of financial returns to the state. First
it yielded on the issue of a purchased shareholding.[42] The Minister made another important concession by reducing the royalty rate from 10 per
cent (the figure he originally sought) to 4.5 per
cent. Department of Industry and Commerce officials
have claimed that this reduction was a trade-off
against Tara's granting to the state its mineral
rights at Navan. Royalties could not, of course,
have been charged on those minerals while they
remained in private ownership. However the mineral
rights involved related to only 6 million tons of
ore, and a simple numerical calculation will quickly
show that the Minister was very much the loser by
this bargain.

Turning to the conduct of negotiations, a
number of points emerge. First, the perceptions of
the parties involved of one another played a central
role in shaping events. Tara was convinced that
Keating would not compromise on the basic issues
involved in the negotiations, and in consequence
adopted an aggressive strategy apparently designed
to force his hand. Keating eventually made substantial financial concessions, but claims that Tara
could have won them without confrontation. It is
apparent that, had Tara based its approach on the
experience of international mining companies in
other countries, it would not have accepted
Keating's intransigent attitude in the early negotiating sessions at its face value.

On his part, Keating mistakenly took for
granted that Tara understood the political imperatives, ideological and practical, which led him to
stand fast on the issue of state participation. In
consequence, the Minister misinterpreted Tara's
reaction to his stand, attributing that reaction to
influence exercised by Tara's foreign owners. This
lack of understanding on both sides led to the
adoption of entrenched positions, an outcome which
imposed costs on all involved.

A second point is that Tara made an error of judgement if it decided that it could successfully 'take on' the Irish government. Certainly the legal and political constraints arising from provisions of the Irish Constitution protected the company from retaliation, but Tara may possibly have under-estimated the cohesion of the Coalition government and Keating's ability to 'sit tight' while pressures mounted on the company. Those pressures did mount, and eventually forced Tara to compromise. A number of commentators have noted that, as a general rule, private corporations operate under much tighter time constraints than do government agencies.[43] Tara apparently ignored this rule, or at least believed that it could prove an exception to it, and the company suffered as a result.[44]

Third, Keating introduced an extraneous factor into the bargaining situation with Tara. The Minister was prepared to make financial concessions to the company in order to establish principles which would be applied to off-shore oil operators. It seems probable that the Minister could have secured a higher financial return to the state from the Navan development had he been prepared to make concessions in the area of state participation. The question arises whether he should not have attempted to obtain as high a return as possible from minerals known to exist, rather than concern himself with hypothetical oil discoveries. Only time will tell whether his decision was justified.[45]

Fourth, the Irish government's 'objective' bargaining power was not fully mobilised or util-ised, but in general this did not result from any pay-offs to or trade-offs with foreign investors. The following factors were responsible. (1) The relevant technical and financial expertise could not be fully developed because of institutional and bureaucratic constraints. (2) The Irish Constitut-ional framework imposed limitations on government action. The Irish authorities did not believe that it was desirable or possible to change that frame-work because they perceived in the Irish electorate (and themselves felt) an adherence to certain norms, the most significant of which involved a belief in the 'sanctity' of private property. Those norms existed quite independently of any relationships between foreign investors and domestic political groupings. (3) The position of the relevant govern-ment Minister was undermined by lack of support among his party colleagues, a lack of support which in general was apparently unconnected to any

The Navan Zinc/Lead Development

relationship between themselves and the foreign
investors concerned.

Notes

1.  <u>Irish</u> <u>Times</u>, 15 February 1975.
2.  Interview, Radio Eireann, 4 July 1974.
3.  The following section is based largely on an
extensive interview held with Mr Keating in Dublin
on 1 December 1978. Unless otherwise stated, all
quotations are taken from this interview.
4.  Interview, <u>Management</u>, December 1973, p. 15.
5.  <u>Irish</u> <u>Times</u>, 14 August 1973.
6.  Interview, 1 December 1978.
7.  Ibid.
8.  <u>Irish</u> <u>Times</u>, 26 September 1973.
9.  The controversy which followed the ending of the
twenty year exemption has been described in detail
elsewhere. See C. O'Faircheallaigh, 'The Role of
Foreign Investment in the Development of the Irish
Mining Industry, 1955-1975', unpublished MA dissertation, Australian National University, 1975, pp.
57-64.
10. O'Kelly, 'Legislation and the Irish Mining
Industry', p. 153.
11. <u>Dail</u> <u>Debates</u>, 1 May 1974.
12. Derived from Table 2.1 and n. 37, p. 69.
13. Interview, 1 December 1978. Correspondence with
Mr Keating, 1 November 1982.
14. Interview, 1 December 1978.
15. Most of those companies operate in Ireland
through wholly-owned subsidiaries, which are private
companies and therefore not required to publish
their accounts under Irish law.
16. The relevant section of the Act is quoted above.
Article 43 of the Irish Constitution, in the following terms, guarantees protection of private
property: 'The State accordingly guarantees to pass
no law attempting to abolish the right of private
ownership or the general right to transfer,
bequeath, and inherit property'.
17. Interview with Department of Industry and Commerce official, 11 February 1978.
18. <u>Irish</u> <u>Times</u>, 30 January 1975.
19. Tara Exploration and Development Company
Limited, <u>Annual</u> <u>Report, 1972</u>, p.3.
20. Interview with Mr Keating, Dublin, 1 December
1978. The practice to date had been to compensate
owners only to the agricultural value of the land
concerned.
21. Keating had hoped, for example, to employ acc-

93

ountants and economists with experience of working in the mining industry. Interview, 1 December 1978.
22. Interview with Tara Mines Ltd. Public Relations Officer, Dublin, 12 February 1976; McEwan Securities Ltd., Tara Exploration and Development Company Limited, Toronto, 1973.
23. Cominco's bid placed a value of C$32.4 million on a 20 per cent shareholding in Tara, and as original shareholders the Hughes and associated interests would have paid only a fraction of the bid price for much of their shareholding.
24. Irish Times, 14 November 1975, 2 March 1978.
25. Interview with Mr Keating, Dublin, 1 December 1978.
26. Interview with Department of Industry and Commerce official, Dublin, 11 February 1976; interview with Mr Keating, Dublin, 1 December 1978. Tara subsequently issued a court writ which sought, inter alia, a ruling that the Minister would be going beyond his powers in demanding equity participation in the Company (see below).
27. In dealing with the wider context of negotiations between international corporations and host governments, A. Kapoor has noted that governments frequently attempt to set precedents which may eventually lead to the achievement of particular policy aims, even if they are required to make concessions in other areas at that particular point in time. See A. Kapoor, 'International Business – Government Negotiations in Developing Countries', in A. Kapoor and P.D. Grub (eds), The Multinational Enterprise in Transition (The Darwin Press, New Jersey, 1973), pp. 367–83.
28. Irish Times, 24 August 1974.
29. Interview with Tara Mines Limited Public Relations Officer, Dublin, 12 February 1976; see also the statement by Tara's counsel in the Supreme Court, reported in Irish Times, 31 January 1975.
30. Interview, 1 December 1978.
31. For example, Litvak and Maule have described how the Guyanese government, in bargaining with foreign bauxite-mining companies, put forward its initial position as 'non-negotiable' when it was in fact willing to compromise on major issues. See I.A. Litvak and G.J. Maule, 'Forced Divestment in the Caribbean', International Journal, Vol.XXXII, No. 3 (Summer 1977), p.505.
32. Interview with Tara Mines Limited's Public Relations Officer, Dublin, 12 February 1976.
33. Irish Times, 3 September 1974.
34. Interview with Mr Keating, Dublin, 1 December

1978.
35. Irish Times, 31 August 1974.
36. See, for example, Irish Times, 31 August 1974.
37. Interview with Mr Keating, Dublin, 1 December 1978.
38. Ibid.
39. Ibid.
40. Irish Times, 14 November 1975, 2 March 1978.
41. Tara Mines Ltd's output is sold to independent smelters while its parent does not produce, or supply it with, mining inputs.
42. It is impossible to express this concession in monetary terms, as one cannot predict what Tara's profit margins will be.
43. See, for example, Kapoor in Kapoor and Grub, The Multinational Enterprise, p. 375.
44. It is uncertain how much the closure actually cost Tara. However, given that inflation in Ireland was then running at an annual rate in excess of 20 per cent and that interest charges on existing loans would have been substantial, the cost was certainly high.
45. As this book is going to press, Ireland has just announced its first (apparently) commercial oil find.

Chapter Four

FOREIGN INVESTMENT AND THE ZAMBIAN COPPER INDUSTRY,
1964-1970

Historical Introduction

In 1891 the British government granted control of an
indeterminate region north of the Zambezi river to
the British South Africa Company (BSAC), formed two
years earlier by Cecil Rhodes. The territorial
limits of the grant were subsequently fixed (with
little reference to existing linguistic or tribal
boundaries) by Britain, Germany, Portugal and
Belgium; this area, which later became Zambia, was
then inhabited by some 100 distinct tribal groups,
and covered 285,000 land-locked square miles of the
Central African Plateau.[1] It was governed (as
Northern Rhodesia) by the BSAC itself until 1924,
when the British Colonial Office assumed
administrative control. A legislative council was
then created; because Northern Rhodesia's electoral
laws disenfranchised Africans, it was dominated by
representatives of the white settler community.
Though the Colonial Office was officially
responsible for the administration of Northern
Rhodesia, the legislative council exercised a major
impact on government policy.
     The BSAC's interest in the country derived from
its potential mineral wealth. From 1923 onwards, it
granted exclusive prospecting rights over extensive
areas to a number of mining companies, many of which
were interlocked in a complex web of mutual or
shared ownership,[2] at the centre of which stood the
Anglo American Corporation of South Africa (AAC),
and the US-based Selection Trust. During the 1920s,
these companies discovered a number of massive sul-
phide copper deposits within an area 90 miles long
by 30 miles wide, known since as the Copperbelt.
These orebodies were developed by foreign capital
and imported skilled labour. The first of the large

mines opened in 1931, and others quickly followed. Expansion continued during the 1940s and 1950s in response to increases in the Western World's demand for copper. By 1964 Zambia's annual production had reached nearly 630,000 tonnes, making her the world's third largest copper producer[3].

In 1953 Northern Rhodesia joined with Southern Rhodesia and Nyasaland to form the Federation of Rhodesia and Nyasaland. The Federal government, based in Salisbury, was firmly under the control of Southern Rhodesian settlers, who used their political power to concentrate economic development in the south. Customs barriers were removed within the union, and industries created to take advantage of the regional common market were almost exclusively located in Southern Rhodesia.

Federation was deeply resented by Africans in Northern Rhodesia; against their will they had become subjects of a government which sought to perpetuate white domination and racial discrimination. The African National Congress (ANC) was established to oppose Federation. In 1958 the movement split and its more militant faction, led by Kenneth Kaunda, formed the United National Independence Party (UNIP). Universal suffrage was introduced in 1963, and in January 1964 UNIP won a convincing electoral victory. Kaunda requested immediate independence from Britain, and the Republic of Zambia came into existence in October 1964.

The development of the copper industry had exercised little direct impact on Zambia's predominantly subsistence economy. A high proportion of the industrial goods and services used by the mines were imported;[4] few linkage industries emerged, and those which did (power generation, rail transportation, timber and metal products) failed to achieve self-sustaining growth. Because of the Copperbelt's geographical concentration, mine infrastructure did little to stimulate other economic activities. Neither did wage payments represent a significant stimulant – African wages were low (especially until 1953), while expatriate workers remitted a large part of their incomes or spent them on imported goods.

Neither were mining company profits or mineral taxes expended so as to stimulate eocnomic development on a broader front. In general, the mining companies did not diversify their activities, remitting their profits or using them to expand copper production. The BSAC made no substantial investments in Northern Rhodesia over the period 1930-63, during which time it received some £90 million in net profits from its mineral rights. AAC and Rhodesian Selection Trust (RST) reinvested part of their profits in mining activities, but the rest they remitted; over the years 1954-63, for instance, dividend payments by the various operating companies amounted to £259 million.[5] Mineral taxation levels were low until 1948, and from 1953 onwards a substantial portion of mineral revenues was channeled to Southern Rhodesia and Nyasaland.[6] Much of the public expenditure which was made in Northern Rhodesia was directed towards the monetary sector, while little reached the rural areas away from the line-of-rail running from Livingstone in the South through the Copperbelt to the Katangan border. In general, patterns of public expenditure discrim- inated heavily against Africans; for example, during the federal period expenditure on European education averaged £103 per child, on African education £9 per child.[7]

Economic discrimination characterised the private as well as the public sector. Fertile agricultural land along the line-of-rail was reserved for Europeans, as was a large share of the profitable local maize market. On the Copperbelt, European mineworkers established, and for many years maintained, a monopoly of all skilled positions. Racial discrimination took social as well as political and economic forms, and many of those who formed Zambia's first government, including the President, Dr Kaunda, had suffered personal humil- iation under a system of racial segregation which branded them as second class citizens in their own country.[8] It is important to keep this background of discriminatioin and deprivation in mind, for it was to exert a strong influence on government policy towards the mining companies in the period after Independence.

Taxation and the Formulation of Mineral Policy 1964-9

At Independence, Zambia's economy was characterised by four notable features, each a cause of considerable concern to the new government. The most apparent was the sharp discrepancy between a rural subsistence sector in which some 70 per cent of Zambia's four million people eked out a precarious existence, and an urban sector clustered along the line-of-rail, based on a large modern mining industry and providing a small number of expatriates (about 78,000 in 1959) with a standard of living high even by north European standards. Some 225,000 Africans were also employed in the urban sector, at incomes far below those of expatriates but well above those of subsistence farmers.

Another important feature was the degree to which the economy was dependent on copper mining. In the early 1960s, it provided about 90 per cent of exports, from 35 to 70 per cent of government revenue (depending on copper prices), and accounted for about 48 per cent of net domestic product.[9] The corollary of this dependence was the relative insignificance of manufacturing and agriculture - in 1964, mining accounted for 47 per cent of GDP, manufacturing for only 6 per cent, and agriculture for 11 per cent.[10]

A third feature involved the almost total absence of an indigenous skilled workforce, reflecting the neglect of African education and the lack of training facilities for Africans on the mines. (It was illegal for an African to become an apprentice before 1959.) At Independence there were only 103 Zambian university graduates and 1,200 high school graduates, while the mining industry's skilled workforce was comprised almost exclusively of expatriates. The fourth feature was Zambia's dependence on her white-dominated southern neighbours, Rhodesia and South Africa. Rail links passing through Rhodesia carried about 90 per cent of Zambia's exports and imports, while in 1964 Rhodesia supplied about 40 per cent of Zambia's imports and South Africa a further 20 per cent.[11]

The structure of the economy could not be altered without massive government investment over a substantial period of time and, given Zambia's economic circumstances, much of the required investment would have to be financed from mineral revenues. Thus it was crucial that the Zambian government develop a mineral taxation system which would produce substantial revenues without discouraging continued investment in the copper industry; its first initiative on mineral taxation seemed to augur well for its ability to do so.

The BSAC had retained the right to impose mineral royalties when it granted leases to AAC and RST in the 1920s. Royalty payments equalled 13.5 per cent of the 'price' of copper (in fact, the average London Metal Exchange (LME) price over a given period), less K16,[12] per long ton produced. In other words, royalties became payable once the LME price exceeded K118.52 per long ton. By the late 1950s, prices were well above this level and royalty payments were consequently substantial, amounting to nearly K24 million in 1960. The Zambian government wished to capture the BSAC's mineral rights and revenues. It was able to demonstrate that the legal basis of the BSAC's ownership of mineral rights in Zambia was extremely tenuous, and consequently to argue that the Company was not entitled to any compensation if and when it surrendered those rights. Nevertheless, anxious to assure the international financial community of Zambia's good faith, it offered the BSAC nominal compensation of K4 million, an offer which was eventually accepted.[13] Thus it showed an awareness of the need to reconcile short-term revenue requirements with Zambia's longer-term need for continued investment.

However, in the years after 1964 mineral taxation became an increasingly contentious issue in Zambia. Two types of taxation were involved. Royalties were paid by all mines according to a single formula (outlined above), while income tax was charged on profits after deduction of royalty payments. The maximum income tax rate was 45 per cent, but lower rates applied to companies mining low grade ores or engaged in expanding production.[14] The mining companies opposed the imposition of royalties, claiming they might make it unprofitable to extract lower-grade ores from existing mines, since they were a charge on production which took no account of costs. Royalties might also deter investment in new mines by reducing or eliminating potential profits. Any tax would of course reduce profits, but a royalty (unlike a profits tax) could eliminate profit altogether, or even turn it into a loss. The likelihood of this happening was increased by the fact that Zambian copper was then sold at producer prices which were often substantially lower than the LME prices on which royalties were calculated. The companies also argued that application of a single royalty rate to all mines was inequitable, since less profitable companies paid higher effective tax rates than more profitable ones.[15]

Initially, it appeared the government would amend the tax system,[16] but in the event no initiative was taken. Government spokesmen justified their inaction on the grounds that the system had worked well for the BSAC and had provoked little criticism from the companies prior to Independence, and consequently that it should work well for

Zambia.[17] Zambian politicians were understandably
suspicious of the companies' new-found concern with
the royalty system, but their suspicions diverted
their attention from the fact that significant
changes were underway in Zambia's mining industry.
The most important involved the cost structure of
copper extraction. Ore grades were falling,
requiring increased throughput to maintain metal
production. Mining was taking place at deeper
levels, and consequently ventilation, drainage, and
haulage costs were increasing. Labour costs were
also rising rapidly.[18] Over the period 1964/5-
1967/8, the total average cost per ton incurred at
the mine increased by 45 per cent.[19] At the same
time, transport costs were rising, particularly
after the re-orientation of trade routes which
followed the Rhodesian Unilateral Declaration of
Independence (UDI) in 1965.[20] Profits were already
being eroded, and royalties consequently represented
a proportionately greater burden. That burden might
be bearable while prices were high, but when prices
fell it was bound to have a major impact.

Another significant change concerned the
attitude of the copper companies themselves. From
the mid 1930s to the late 1950s, AAC and RST had
been attempting to increase their share of the world
copper market, a strategy which called for major
expansion of mining capacity in Zambia. By Independence, this situation had changed. As one
commentator noted in 1964, the companies 'are
earning a comfortable return ... under present
market-sharing arrangements, and apparently will not
attempt to enlarge their market share in the
future'.[21] Political developments had also
influenced corporate attitudes. Company officials
feared an independent government might nationalise
the mines, force localisation too quickly, raise tax
rates, or allow political considerations to
influence marketing policies. Their inclination was
to avoid major new commitments until the character
of the new regime became clear, though sufficient
funds would be invested to maintain the value of
their assets.[22] In sum, whereas the companies had a
strong inclination to invest in new capacity during
the decades prior to Independence, they were now
disinclined to do so, and positive incentives were
required if their attitude was to be changed. The
tax system in force in 1964 certainly did not offer
such incentives.

In April 1966 the Zambian government introduced
another tax, the copper export tax, to capture part

of the additional revenues accruing to the copper companies as a result of their switch from a producer price system to an LME-based system.[23] This tax would equal 40 per cent of the amount by which the LME price exceeded £300 (K600) a ton. The copper export tax succeeded in its main aim, earning K245 million for the government over the following three years,[24] but it compounded the disadvantages of the royalty system since it too was a charge on production which took no account of cost.

Despite its inconsistencies and its weaknesses, Zambia's mineral taxation system remained unchanged until 1970. The government's lack of initiative in this area reflected its failure to come to grips with the basic problems of formulating a mineral policy to suit Zambia's circumstances. Public policy statements invariably emphasised the need to increase production,[25] but the implications of this requirement were not examined.[26] Would it be necessary to sacrifice tax revenue in order to encourage investment in new capacity? If so, how much was the government prepared to forego? In fact, was the priority to maximise production (and so employment and export receipts) or was it to maximise revenue?

What explains this failure to formulate clear policy priorities for an industry so crucial to Zambia's economy? The importance of copper mining may itself provide part of the explanation. Zambia's political leaders were conscious of their ignorance of the industry,[27] and may have been reluctant to take initiatives which might damage it, especially when existing arrangements were producing very substantial revenues. In addition, the government was not under significant political pressure to act. The parliamentary opposition made only feeble attempts to capitalise on the government's inaction,[28] while vigorous political debate was generally lacking in a country where the politically able were nearly all employed by government or the public service. Zambia's leaders faced other major problems which were of immediate political significance (the Rhodesian situation, rural development), and may have been reluctant to utilise the limited administrative capacity at their disposal to formulate policies for an industry which was then operating very profitably.

Probably as a result of these factors, appropriate machinery for formulating mineral policy was not established. A separate Mines Department was not created, and Mines shared ministries in turn with

Labour, Cooperatives, Industry and Trade, and Lands;
thus the attention of the politicians responsible
for mining was divided. In 1968 a Mines and Minerals
Development Unit was established within the Ministry
of Lands and Mines, and this might have formed the
basis for efficient policy-making. However its staff
was drawn from very disparate backgrounds, their
individual capabilities varied markedly, turnover
was high, and consequently the quality and content
of advice given to policy-makers varied considerably
over time. For this reason, and because its expatriate staff was distrusted by some politicians, the
unit had little impact on policy-making[29]. Turnover
at the ministerial and permanent secretary level was
also high,[30] making it difficult for the individuals
involved to acquire relevant expertise, and militating against continuity in policy-making.
The weakness in policy formulation, and
especially the failure to tackle the problems
created by Zambia's mineral taxation system,
reinforced the reluctance of the copper companies to
undertake major new investments in Zambia. AAC and
RST were waiting for indications of the government's
competence and of a favourable attitude towards
themselves, and they regarded the royalty issue as
an important test. They were not asking that their
overall tax burden be significantly reduced, but
rather that it be redistributed from less profitable
to more profitable mines, and it would therefore
cost the government little to grant what they
regarded as a major concession. The government
failed the test, with the result that the companies
(particularly RST) deferred consideration of a
number of major new mineral developments.[31]

The Take-over Decision

Over the period 1964-8, the Zambian government
became increasingly dissatisfied with the pricing,
dividend, credit and investment policies of the
foreign companies which dominated Zambia's
economy.[32] In April 1968, in a speech at Mulungushi,
President Kaunda announced a series of measures
designed to give the government greater control over
their operations. These included curtailment of
local borrowing by the companies and a limitation on
dividend payments to the lower of 50 per cent of
profits or 30 per cent of equity capital. In
addition, the government would greatly expand its
direct involvement in the economy, by asking a
number of major manufacturing and distribution com-

panies to 'invite' the state to purchase 51 per cent of their equity.[33] The vehicle for state particip- ation would be the Industrial Development Corpor- ation (INDECO). INDECO would negotiate values and terms of payment for the state's shareholdings with the companies involved, but the President stated that compensation would be based on book value.[34]
No mention was made of government participation in the mining companies, but the President did express criticism of their investment policies:

> I want to say to the mining companies that I am very dissapointed at the virtual lack of mining development since Independence... The companies claim that the royalty system has been against new development ... I do not agree ... I have been following their accounts and I know very well that they could have embarked upon further expansion if they chose to devote part of their profits for this purpose. Instead of re- investing they have been distributing over 80% of their profits every year as dividends.[35]

Nevertheless, Dr Kaunda stated, the government would change the royalty system. No details were announ- ced, but the new arrangement would leave government revenue substantially unchanged while meeting the companies' criticisms. The companies for their part would be expected to expand production; they would be subject to the general limitation on dividend remittances and would be expected to reinvest the 50 per cent of profits retained in Zambia.
Was this criticism of the companies justified? First, it should be noted that two separate issues were in fact involved - development of additional capacity and reinvestment of profits. The rate of development of new capacity certainly slowed down after 1964. Copperbelt production increased by only 14 per cent between 1964 and 1969 (see Table 4.1), campared with 21 per cent between 1960 and 1965 and 72 per cent between 1955 and 1960.[36] However after Independence mining was taking place at deeper levels and ore grades were declining, and the amount of investment required to maintain production increased. Investment of this type was charged to working costs rather than appropriated from profits (a more usual procedure), and for this reason Pres- ident Kaunda's claim that only 20 per cent of profits were reinvested was somewhat misleading. For example, AAC group mines declared profits of K101.4 million over the period 1964-8, of which K12.6

Table 4.1: Zambia's Copper Production, Selected Years 1942-64, and 1964-79 (000s Metric Tonnes)

| Year | Production | Year | Production |
|------|-----------|------|-----------|
| 1942 | 251 | 1969 | 720 |
| 1946 | 185 | 1970 | 684 |
| 1950 | 281 | 1971 | 651 |
| 1954 | 385 | 1972 | 718 |
| 1958 | 380 | 1973 | 707 |
| 1960 | 589 | 1974 | 698 |
| 1964 | 632 | 1975 | 677 |
| 1965 | 696 | 1976 | 709 |
| 1966 | 623 | 1977 | 656 |
| 1967 | 663 | 1978 | 643 |
| 1968 | 685 | 1979 | 588 |

Source: Baldwin, Economic Development and Export Growth, Table 2.2, p. 33; Zambia Mining Year Book, various issues.

million or 14 per cent was appropriated for capital expenditure. But capital investments totalling K36.4 million were made in maintaining production,[37] and if these had been appropriated from profits rather than charged against working costs AAC's reinvestment rate would have been 36 per cent. This is similar to the rate achieved by the Northgate group in Ireland (37 per cent),[38] and by Bougainville Copper Limited in Papua New Guinea (41 per cent).[39] Significantly, neither of these companies expanded capacity substantially during the periods to which these figures apply.

This discussion illustrates a point made earlier: the Zambian mining groups were willing to maintain the value of their investments, but were reluctant to invest in new capacity (for reasons mentioned above). The measures announced by President Kaunda did nothing to overcome this

reluctance, first because his promise to change the
royalty system was not acted upon, second because
restriction of dividend remittances had no impact on
investment. The companies simply accumulated the
blocked funds in local bank accounts.

The issue of prospecting and mining rights also
required urgent attention if new mineral developments were to proceed. The government had acquired
only the BSAC's mineral rights, and other companies
(including AAC and RST) held rights, in perpetuity,
over a large part of the country. If the companies
decided to 'sit' on known mineral deposits or to
ignore certain areas, the government could neither
force them into activity or award mining or prospecting rights to other, more energetic firms.
Corporate ownership of mineral rights also prevented
the state from exercising control over aspects of
mining operations considered vital to its interests,
for example production and recovery rates and cut-
off grades.[40]

By early 1969 it was apparent that the long-
awaited initiative on mineral policy would soon
materialise. Andrew Sardanis, a forceful businessman
who as Chairman of INDECO had implemented the take-
overs announced at Mulungushi, was appointed
Permanent Secretary of the Trade, Industry, and
Mines Department, and drafting of legislation on
taxation and mineral rights was soon under way. By
July the new legislation was ready; its details were
announced by the President at Matero Hall, Lusaka,
on 11 August. Dr Kaunda began by repeating his claim
that AAC and RST were failing to reinvest enough of
their profits in development. He went on to mention
the problems caused by corporate ownership of
mineral rights, and announced that all such rights
would revert to the state. Existing mines would be
granted mining leases for twenty five years, the
size of the lease depending on the rate of
extraction. Where minerals had been discovered but
not exploited, the finder would either proceed with
development within three years or surrender his
rights. Companies engaged in prospecting could complete their work, and if minerals were discovered
apply for mining leases. The rights of companies not
engaged in exploration or development would revert
to the state immediately. The President also
announced that all new mining ventures would be
expected to offer the state at least 51 per cent of
their shares.[41]

Proposed changes to the taxation system were
also outlined. Royalties and export tax would be

replaced by a mineral tax amounting to 51 per cent
of profits. Income tax at the standard rate (45 per
cent) would be charged on the balance of profits,
giving a total tax rate of 73.05 per cent.

These initiatives had been expected, but the
President's third major announcement had not. The
AAC and RST mines would be asked to sell 51 per cent
of their shares to the state. INDECO would negotiate
purchase terms, but again book value would form the
basis of compensation. The state would not pay for
its share immediately, but would do so out of future
dividends.[42]

What prompted the government to seek majority
ownership of the copper mines? The official reason
was that state control of the mines was essential if
economic independence was to be achieved. As President Kaunda stated, '... I wanted the mining
economic reforms to set the Nation well and truly
towards its economic independence. I do not think
that this can be achieved without the Nation
acquiring full control of the existing mines...'[43]
The belief in the need for state control was based
on the assumption that national and corporate
policies would frequently conflict. Six specific
areas of potential conflict were indentified,[44] the
first of which involved the mining companies' purchasing policies. Zambia was determined to reduce
imports from Rhodesia/South Africa (which traditionally supplied a high proportion of mining
industry inputs), even if alternative sources of
supply were more expensive.[45] The mining companies,
on the other hand, wished to stick to suppliers whom
they knew well (indeed some were affiliated
companies), and on whom they could therefore rely to
meet their requirements efficiently. In addition, it
was usually cheaper and almost always quicker to
obtain supplies from Southern Africa than from
Europe or elsewhere.

A second area, that of investment policies, has
already been discussed in detail. A third involved
the issue of Zambianisation. Until shortly before
Independence, nearly all skilled and supervisory
positions in the mining industry were held by
Europeans, many of whom regarded themselves as permanent residents of Northern Rhodesia. Over the
period 1963-9, important advances were made in
Zambianising semi-skilled and lower supervisory
positions, and all Europeans were placed on fixed-
term contracts. But progress had been slow in the
skilled manual, technical, professional and
managerial categories. For example, in 1969 over 95

per cent of artisans were still Europeans, while
there were no Zambian underground managers,
assistant underground managers, mine captains,
metallurgists, or mining engineers.[46] A fourth
concern was that the companies might employ methods
such as over-invoicing of goods purchased from
affiliates to remove profits from Zambian
jurisdiction. Fifth, it was feared that mining
methods might be employed which were inimical to
Zambia's long-term interests, for example in the
areas of grade control and dilution. Finally, it was
felt that employment, wage, and pricing policies
adopted by the companies might conflict with
national economic and social goals. In addition to
and partly arising from these specific issues, there
was a general feeling that Zambia's sovereignty
would be undermined if an industry on which its
future depended was controlled by foreigners.

Two important issues arise: were these fears
justified and, if so, would majority state ownership
remedy the situation? These questions are considered
in the next Chapter, but of immediate relevance is
that Zambia's leaders clearly answered both in the
affirmative. Thus the initiatives announced at
Matero can be seen as continuing a broad strategy,
formulated as a logical response to perceived
problems and first implemented at Mulungushi.

However other factors played a part in the
decision to take those initiatives and in the timing
of that decision, and, as we shall see, its timing
was to exert a major impact on the outcome. In 1968
the government had felt that the cost of
compensation on a scale sufficient to persuade the
companies to continue managing the industry would be
prohibitive.[47] But the take-overs announced at
Mulungushi had been paid for out of future profits,
effectively relieving the state of financial cost,
and INDECO had negotiated contracts with the previous owners under which they continued to provide
managerial and technical services. There seemed no
reason why the mining companies could not be taken
over in the same way, especially since the Chileans,
whose expertise was highly regarded in Lusaka, had
recently nationalised their copper industry under a
similar arrangement.[48]

Another relevant point was that major initiatives in mineral policy were about to be taken in
any case, and it seemed sensible to implement all
the desired changes at once, placing the industry on
a firm footing and eliminating uncertainty. But
perhaps the most important influence on the timing

of the decision arose from short-term political considerations. In August 1969 President Kaunda was undergoing the severest political test of his career. Shortly before, a major political crisis had resulted from clashes between the President and the judiciary. UNIP was deeply divided along tribal lines. By August 1969 these divisions had crystallised around Simon Kapwepwe, Vice-President of Zambia and UNIP, and the Matero meeting of UNIP's National Council seemed certain to witness open conflict. The announcement of the latest set of 'reforms' was sure to divert attention from these issues. In the event the President used his considerable powers of oratory to make the most of the occasion, and his success in rallying the Party behind him was almost total. It placed him in such a strong position that he was able to neutralise the threat from Kapwepwe within a few weeks.

Thus the Take-over decision was both a logical outcome of the government's broader economic and political strategy and of shorter-term tactical and political imperatives.

The public reaction of AAC and RST to the Matero announcement was moderate. Neither questioned the basic principle of majority state participation, and both indicated their willingness to negotiate compensation payments and management contracts. Negotiations took place over the period September-November 1969, and an agreement was signed on 24 December 1969, the principal provisions of which are outlined below.[49]

The Terms of the Take-over

Two new mining companies, Roan Consolidated Mines Limited (RCM) and Nchanga Consolidated Copper Mines Limited (NCCM), were formed through amalgamation of the RST and AAC group mines respectively. Ownership would be divided between the government (51 per cent) and the private shareholders.[50] The government would appoint a majority of the directors of both companies, to be known as 'A' directors; the minority shoreholders would elect the remaining, 'B', directors, who would chose the Managing Director of each company from among their number.

AAC would receive K125,766,000 in compensation, RST K84.15 million. In both cases, payment would be made by the Zambia Industrial and Mining Corporation (ZIMCO), though the government's shares would be held by the Mining Development Corporation (MINDECO). ZIMCO would effect payment by issuing

bonds bearing interest at 6 per cent, repayable to
RST in sixteen semi-annual instalments and to AAC in
twenty four semi-annual instalments. Provision was
made for accelerated repayment in certain circumstances; if two-thirds of MINDECO's dividends from
either RCM or NCCM exceeded the guaranteed fixed
annual payments, the additional income would be
applied to bond repayment. If MINDECO's total
dividend income fell below the fixed annual payments, on the other hand, no postponement of those
payments was provided for. Thus the state might not
in fact pay for its shares out of future profits.

A management and consultancy contract was negotiated in both cases. Details of AAC's contract
were not released, but it was similar to RST's,
outlined below. RST would continue to provide the
managerial, financial, commercial, technical and
other services which it had previously supplied to
or procured for the individual mines. These services
would include production planning, capital expenditure estimates, advice on mining operations,
engineering consultancy, recruitment of expatriate
staff, and purchasing of goods outside Zambia. In
exchange for these services, RST would receive 0.75
per cent of the operating company's gross sales
proceeds, 2 per cent of its consolidated profits, an
engineering fee of 3 per cent of the constructed
costs of projects, and a recruiting fee of 15 per
cent of each expatriate employee's first year's
gross earnings.

AAC and RST would continue to market Zambia's
copper output. Their sales contracts would be
exclusive, and remuneration would equal 0.75 per
cent of gross sales. In fulfilling their contracts,
RST and AAC undertook to give training and
experience in relevant functions to staff seconded
from RCM, NCCM, or MINDECO.

The Articles of Association of RCM and NCCM
contained important provisions relating to managerial control of the companies. RCM and NCCM would
pay dividends equal to their consolidated net
profits 'after deduction only of appropriations in
respect of capital expenditure, expenditure for
exploration and prospecting, and reserves for
necessary working capital'.[51] Appropriations for
capital or exploration expenditures would require
the approval of both 'A' and 'B' directors voting
separately, as would other major corporate actions,
for example alteration of RCM's or NCCM's Articles
of Association, participation by RCM or NCCM in non-
mining activities or in mining activities for which

they could not obtain finance on commercially
competitive terms or which the 'B' directors did not
regard as commercially viable, or 'any act ... that,
in the opinion of a majority of the 'B' directors,
[was] not directed towards ... the optimization of
production and profit of RCM and NCCM.' However, the
'B' directors would not 'unreasonably' withhold
their consent 'having regard to the interests of RCM
(and NCCM) and to the interests of the
shareholders'.[52]

The Zambian government agreed to refer any
disputes arising from the arrangement to the International Centre for the Settlement of Investment
Disputes (ICSID). It also agreed that all disputes
would be governed by the law of Zambia as in force
at the time the arrangement was concluded,
disregarding any subsequent legislation. In
addition, the government made specific undertakings
as regards taxation and exchange control legislation. While any part of the ZIMCO bonds were
outstanding, the overall rate of tax payable by RCM
and NCCM would not be increased, and dividends paid
by them would be free of Zambian taxation, as would
all payments arising from the ZIMCO bonds. Provisions relating to capital allowances and to tax
treatment of exploration expenditure would also
remain unchanged.[53] Bond repayments, interest payments, dividends payable to non-Zambian shareholders, and sales and management contract fees
would all be exempt from exchange control.

Breach of any aspect of these arrangements by
the Zambian government (including the management
contracts) would make all the ZIMCO bonds due and
payable immediately.

The Impact of the Take-over

To what extent did these terms allow realisation of
the Zambian government's objectives? On one level,
its aim was to acquire an income-generating asset
free of financial cost, and thus it hoped to pay for
its shareholding out of future profits. In the
event, the ZIMCO bonds were redeemed in September
1973 (see below), and so it is impossible to state
with certainty whether that hope would have been
realised. It is possible to compare what ZIMCO's
liabilities to AAC and RST would have been had the
bonds not been redeemed with the dividends which the
government has actually received from NCCM and
RCM.[54] Such a comparison probably gives an accurate
picture of what would have happened had the Take-

over arrangements run their course, since the
failure of RCM and NCCM to pay dividends after 1974
reflected their inability to generate operating
profits rather than any major change of investment
policies following the bond redemption. It indicates
that the government was mistaken in its belief that
its shareholdings could be paid for out of future
profits, and that it would have had to pay about
K100 million in compensation from other sources.

The government's expectations as to the copper
industry's profitability were disappointed for three
main reasons. First, production costs rose rapidly
over the period 1970-7; costs at the mine increased
by 96 per cent in RCM's case and 84 per cent in
NCCM's, while average transport costs rose by 40 per
cent.[55] Second, the period 1974-8 was characterised
by a prolonged trough in copper prices. Finally,
production levels were adversely affected by bottlenecks in transport routes, shortages of skilled
personnel, and a disastrous cave-in at Mufulira,
RCM's largest mine, in October 1970. In fact
production fell after 1969 (see Table 4.1), whereas
substantial increases had been hoped for. The
Zambian government could not have predicted these
developments, but it could have allowed for unfor-
seen circumstances: depressed prices, rising production costs, and natural disasters were no strangers
to the mining industry. Specifically, the government
could have negotiated a clause allowing for postponement or deceleration of compensation payments in
years of low or no profits,[56] corresponding to the
shareholders' right to obtain accelerated compensation when profits were high. The absence of such a
provision represented a serious weakness in the
Take-over arrangement.

The profitability of the operating companies
was also relevant to the balance of payments impact
of the Take-over. It was felt that a high proportion
of profits would flow out of Zambia if private
ownership continued; on the assumption that profits
would exceed compensation payments, outflow of these
payments would not therefore represent a drain on
the balance of payments.[57] In the event, not only
was that assumption not justified, but the factors
responsible (low prices, production shortfalls,
higher production costs)[58] themselves adversely
affected the balance of payments.[59] Consequently the
outflows of foreign currency created by compensation
payments exercised a proportionately greater impact.

By purchasing equity on the terms it did the
government added to the risks inherent in dependence

on the copper industy – compensation payments would
further reduce government revenue in times of low
prices or other adverse conditions, and would exacerbate the impact of such conditions on the balance
of payments. However, over the years 1964-9 Zambia
had experienced no difficulties in balancing its
external transactions, and by 1969 had accumulated
substantial foreign currency reserves. Its public
revenues had exceeded all current and capital government expenditure in each year except 1968. It
could therefore be argued that the government could
well afford to take such risks if they offered the
prospect of achieving a central policy goal –
national control of the copper industry.

The earlier outline of the Take-over terms
indicates that the Zambian government failed to
achieve that goal, as an examination of their
implications for the areas of potential corporate-
government conflict discussed above reveals. The
position as regards purchasing policies would remain
essentially unchanged, since AAC and RST would
retain responsibility for purchasing outside Zambia.
Neither did the government obtain a significantly
greater degree of control over the crucial area of
investment policies. Changes in accounting procedures implemented in 1970 prevent a detailed comparison of corporate investments before and after
that year,[60] but the general trend is clear. As
mentioned above, over the period 1964-8 the AAC
group mines made total capital expenditures equal to
36 per cent of net profits as defined by accounting
procedures employed after 1970; the equivalent
figure for NCCM in the period 1971-4 was 44 per
cent, by no means a spectacular increase. Even this
increase cannot be attributed solely to the Take-
over. Nearly a quarter of NCCM's capital expenditure
during this period was on the Tailings Leach Plant
at Chingola. This development resulted from the
discovery in 1968/9 by AAC staff of new methods for
recovering copper from mine tailings, at costs which
were less than half the average for the group
mines.[61] Use of the new techniques offered the
prospect of very high returns on investment, a
prospect which AAC would have been unlikely to
ignore, especially given the favourable changes in
Zambia's tax regime. Thus this additional expenditure would probably have occurred regardless of
the Take-over.

Comparable investment figures are not available
for RST, but it is significant that RCM distributed
a higher proportion of its net profits in dividends

over the period 1971-4 than RST did during the years 1964-8 (54 as against 50 per cent),[62] while its reinvestment rate was low even in comparison to AAC's in the period prior to the Take-over (32 as against 36 per cent).[63]

It is evident, therefore, that the Take-over failed to bring about a major increase in the proportion of profits being reinvested. There were two aspects to this failure. The first involved the provision in RCM's and NCCM's Articles of Association calling for distribution of all profits 'after deduction only in respect of capital expenditure, expenditure for exploration and prospecting, and reserves for necessary working capital'. A high dividend level was of course essential if the government was to make its compensation payments; but the outcome was that profits could not be accumulated for investment at a later date. Yet this would have to be done if reinvestment rates were to increase substantially, as some years would elapse between initiation of capital projects and the phase of major capital expenditures.[64]

The level of current investment was nevertheless flexible; however, the management contracts gave the minority shareholders considerable influence in this area. They carried out project evaluation and capital expenditure estimates and were thus responsible for assessing the viability of individual investments. Those considered unviable could be vetoed by the 'B' directors. On the other hand, its powers to appoint directors might have enabled the government to critically analyse such assessments, if the individuals involved were adequately qualified and could draw on relevant expertise from an independent source. As we shall see, these conditions were to be fulfilled, but the reinvestment figures quoted above indicate that the formal constraints were too powerful for this to make very much difference.

Neither did the new arrangement give the government a significantly greater say in Zambianisation policies. No provisions relating to Zambianisation were included, other than that the companies supplying management and other services would provide 'suitable experience' to employees of RCM an NCCM. The Managing Directors of the operating companies would retain complete discretion in employing, promoting, and dismissing staff.

As regards the possibility of over-invoicing, the presence of government directors might increase the flow of relevant information and force manage-

ment to justify its choice of suppliers and the
prices paid for major purchases; once again, the
calibre of the government's appointees would be all-
important. The ability of government directors to
monitor the conduct of mining operations would be
more problematic, since they would be unlikely to
possess the necessary specialised technical
knowledge. It is one thing to question whether an
item has been purchased at the best available price,
quite another to question, for example, whether cut-
off grades being applied in a particular mining
situation are appropriate. However, if the
government developed some expert capacity in this
area the situation might be different. Finally, the
government's position as majority shareholder might
enhance its ability to ensure that corporate wage,
employment, and pricing policies were in line with
national economic and social goals, but the power of
the 'B' directors to veto any action not designed to
optimise profit or production could restrict that
ability. AAC and RST would hardly have refused to
adopt policies which did not affect their profitability in any case, while they could now reject
policies which would have that effect.

In order to obtain its shareholding, the government also temporarily surrendered control in
areas where it had previously exercised jurisdiction, particularly in the field of taxation and
exchange control legislation. In addition, it
allowed its freedom to be circumscribed in a more
general way. If it was in breach of any aspect of
the arrangement, all ZIMCO bonds would become
payable immediately, and the minority shareholders
cold refer the dispute to the ICSID. The ICSID is
supported by international financial institutions
such as the World Bank and the IMF, who might well
insist that Zambia abide by any ICSID ruling as a
condition of granting it loans or other forms of
assistance.

The influence of the minority shareholders
might be less in practice than would appear from an
analysis of the formal arrangements. They would
possess a strong incentive to retain the goodwill of
the Zambian government, both in order to protect
their remaining investments and to ensure that ZIMCO
would honour its financial obligations. Repeated use
of their veto would certainly undermine that goodwill, and it would probably only be used in extreme
circumstances. On the other hand, the ultimate power
of decision-making remained with the foreign
investors, who would also wield significant

influence over the day-to-day decision-making process, because of the legal and other constraints contained in the Take-over arrangements.

## Legislative Initiatives, 1969-70

The Take-over was accompanied by major changes to legislation governing mineral rights and taxation. In general terms, the initiative on mineral rights was clearly appropriate, since it opened up new areas for exploration and ensured that existing right-holders would either pursue serious prospecting programmes or abandon their claims. A substantial number of foreign mining companies expressed interest in the areas released, some undertook major exploration programmes, and two (Agip of Italy and Saarberg Interplan of West Germany) had by 1979 signed agreements for development of uranium deposits.

A more specific issue involves the requirement on applicants for prospecting licences to offer the state a 51 per cent share in any mining venture which might eventuate. This requirement represents a basic element of government policy which some companies may find unacceptable, but in addition the manner in which it is framed may itself act as a disincentive to exploration. The state decides whether it will exercise its option when and if a mining lease is applied for; if it does so, it then pays its full share of prospecting and exploration costs. But if the prospecting effort fails, the state incurs no liability; it avoids the initial risk, but benefits from any successes. In justifying this approach, the government argues that it should not invest Zambia's limited financial resources in high risk activities and should consequently leave prospecting to private companies.[65] This argument may be valid in theory, but in practice it may not be possible to attract private investors unless the state accepts some of the initial risk, and in fact MINDECO has become directly involved in exploration ventures on a number of occasions.[66]

Before examining the 1970 tax legislation, two points should be made. First, despite the Take-over, it was hoped that private investment would continue to play a major role in mineral development. Second, the operating companies would continue to function on commercial lines, and would assess potential investments accordingly. In other words, investment decisions crucial to the mining industry would continue to reflect perceptions of likely

profitability, which in turn would be influenced by
the extent and character of taxation.

Under the new tax regime, all taxes are based
on profits, and this removes the more obvious
anomalies of the old system; less profitable mines
no longer pay higher effective tax rates than more
profitable ones, taxes cannot exceed profits, and
unprofitable mines cannot be taxed. The 'reforms' of
1969-70 also removed a great deal of the uncertainty
which had surrounded the whole area of taxation, an
important prerequisite if investment was to expand.
However, the new system has one important drawback.

Application of such a high flat rate of tax
(73.05 per cent) will certainly act as a disincentive to private mining investment. The new rate is
almost equal to the marginal rate under the old
system (74.42 per cent for rises in the copper price
above K600 per long ton), and substantially higher
than the average rate applied over the period since
Independence (62 per cent for the AAC group
mines).[67] The 1970 legislation did allow investors
to recoup their initial outlays before tax was
applied,[68] but because of the high tax rate it would
probably take a considerable period of time for
those outlays to generate a significant return.
Given the relatively short term horizons of
investors in Central Africa, the deterrent effect of
such a system must be significant. That effect would
be proportionately greater where profits were in any
case expected to be slim because of factors such as
low ore grades or geographical isolation, a point
which might apply equally to new mines or to certain
parts of existing ones. The government might
logically disregard the likely effects of the tax
system on marginally profitable operations if its
only concern was with revenue generation. It is in
fact anxious to see such operations proceed because
of the direct economic benefits they are expected to
create (employment, export receipts, infrastructural
development),[69] and an appropriate tax system should
clearly take this requirement into account. In
theory the new system does so, in two ways.

First, mineral tax (but not income tax) can be
reclaimed to the extent that the return on equity
falls short of 12 per cent. In practice, this provision is unlikely to have a significant impact on
investment decisions, as the rate of return it
'protects' is unlikely to approach that required by
foreign investors, especially since major mineral
developments in Zambia are likely to have a fairly
high debt : equity ratio.[70] Second, the Zambian

government stated that it would negotiate lower tax rates for specific projects which would otherwise not be viable. While they might encourage development of marginal prospects which were discovered, discretionary provisions of this type are unlikely to have a significant impact on companies contemplating exploration, since they would not know in advance how eligibility would be determined or precisely what concessions would be available.

In sum, the revised tax system ensures that the Zambian government will continue to reap substantial benefits from existing, profitable mines but, though it removes some major faults of the old system, it continues to deter investment in new projects (particularly marginal ones), in working of lower-grade ores in existing mines, and particularly in exploration. It does not appear to reflect any serious attempt to reconcile Zambia's need for revenue with its need for continued exploration and investment.

This judgement may seem harsh, since significant exploration has taken place under the new system, and since a number of potential projects are now being assessed. However, a closer examination of the companies involved indicates that their investment decisions may be influenced by considerations which are not strictly commerical. Three of the most active groups, including both the companies which have signed development agreements, are state agencies of European countries (France, Germany and Rumania). All three have been seeking uranium, and their involvement in Zambia has been motivated by the desire of their home governments to secure reliable supplies of energy minerals rather than by expectations of profitable investment.[71] One of the most active private groups, AAC/De Beers, holds a monopoly position in the world diamond industry; an AAC representative informed the author that its search for diamonds in Zambia was motivated primarily by its determination to preserve that monopoly, and that it would not wish to develop any discoveries it might make under the existing fiscal regime.[72] It seems therefore that few foreign investors find Zambia an attractive location on purely commercial grounds, yet it is on these grounds that investment in copper and other base metals must be sought.

The Zambian Government's Negotiating Effort

The initiatives of 1969-70 failed to resolve either

of the major issues which concerned the Zambian government - control of the mining industry and taxation and investment. In addition, there were significant drawbacks to the manner in which the government obtained its majority shareholding. Major weaknesses clearly still existed in the formulation and implementation of mineral policies. Some of these resulted from the timing of the Take-over. As mentioned above, short-term political consideration hastened its announcement. Very little detailed preparation had been undertaken beforehand, and only some half a dozen of Dr Kaunda's closest associates even knew that a take-over was planned.[73] Once the announcement was made, the government was anxious to conclude negotiations rapidly to prevent uncertainty among investors and expatriate mineworkers; the whole process of preparation and negotiation took place over a period of only ten weeks.[74]

As a result there was little opportunity to identify policy aims, establish priorities, and select policies designed to achieve their implementation; indeed there was not even time to develop institutional structures within which these processes could take place, or to assemble a body of individuals capable of carrying them out. The government's negotiators were consequently at a serious disadvantage. The failure to establish priorities meant that they were pursuing a wide variety of aims, not all of them mutually consistent and relating, for example, to revenue generation, investment policies and production rates, corporate decision-making, and the extent and character of compensation. They were also at a tactical disadvantage since they had little opportunity to develop their own bargaining positions, and had to react to those of the companies.

The time constraint does not explain the failure to devise a more appropriate tax system, since the new legislation was prepared well in advance. Clearly, Zambia's politicians were not yet convinced of the need to devise workable tax legislation which would encourage mining investment, probably because they believed that the copper industry would continue to operate profitably and to expand production while subject to high levels of taxation, as it had done under the colonial regime. This belief might have been dispelled by a competent advisory staff whose advice was taken seriously, but such a group did not exist.

The efficacy of the advisory staff was of course important to the whole negotiating effort, and it is apparent that important deficiencies existed in this area. As in the earlier period, the staff was diverse in origin (it included Britons, Chileans and Eastern Europeans), and the perspectives and assumptions of its members frequently differed, as did the advice they offered. Staff members did not enjoy close liaison with senior politicians, partly because there was insufficient time either for suitable institutional structures to be established or for mutual confidence to develop. Racial differences may also have created a barrier; many senior politicians were still reluctant to trust white advisers, particularly when dealing with European-owned mining companies.[75] In this context, the absence of suitably-qualified Zambian advisers takes on added significance. That absence resulted partly from a general scarcity of educated personnel, but it also reflected the government's failure to allow senior administrative staff greater stability of tenure, or to devise an effective institutional framework within which they could operate.

Thus the government lacked either the opportunity or the capacity to formulate and pursue policies appropriate to Zambia's circumstances. In consequence it tended either to follow the lead of other LDC copper producers or simply to opt for the status quo. Its decision to seek control of the industry by acquiring majority ownership, for example, was influenced by the earlier Chilean decision to do so.[76] Yet Chile's circumstances were very different - a high proportion of technical and managerial positions in its copper industry were held by Chileans, while its government had developed a wealth of expertise regarding the industry. In the field of taxation, the Zambian government opted for the status quo in its crudest form - it simply enshrined the marginal tax rate which had prevailed over a short (and atypical) period prior to the take-over.[77]

Notes

1. Zambia is bordered by Angola, Zaire, Tanzania, Mozambique, Zimbabwe, and the Caprivi Strip of Namibia.
2. The complexities of corporate ownership have been disentangled by F.L. Coleman, The Northern Rhodesia Copperbelt, 1899-1962: Technological development up to the end of the Central African Federation (Manchester University Press, 1971), Chapters 2, 3

and 4.
3. Zambia Mining Year Book, 1969, p.27.
4. Practically all capital goods were imported, while as late as 1960 only 15-20 per cent of operational stores were purchased locally. R.E. Baldwin, Economic Development and Export Growth: A Study of Northern Rhodesia 1920-1960 (University of California Press, 1966), p. 37.
5. M. Nziramasanga, 'The Copper Export Sector and the Zambian Economy', unpublished PhD dissertation, Stanford University, 1973, p. 27.
6. A. Martin, Minding Their Own Business: Zambia's Struggle Against Western Control (Hutchinson, London, 1972), p. 36.
7. R. Hall, Zambia (Pall Mall Press, London, 1965), p. 172.
8. For Dr Kaunda's own account of his experiences, see his Zambia Shall Be Free - An Autobiography (Heinemann, London, 1972).
9. Zambia Mining Year Book, 1969, p. 28.
10. Derived from C. Elliott (ed.), Constraints on the Economic Development of Zambia (Oxford University Press, Nairobi, 1971), Table 1.9, p.16.
11. Derived from S. Goodman, 'The Foreign Exchange Constraint', in Elliott, Constraints on the Economic Development of Zambia, Table 8.2, p. 245.
12. Zambia established a separate currency unit in January 1968, the Kwacha (K), at the rate of £1 sterling = 2 Kwacha.
13. The transfer of the BSAC's mineral rights has been documented elsewhere. See Martin, Minding Their Own Business, pp. 128-32.
14. In 1965, for example, five different rates were charged (35, 41.25, 42.5, 43.75 and 45 per cent).
15. Martin, Minding Their Own Business, pp. 128-9.
16. Zambia Hansard, 15 December 1964.
17. See the statement by the Minister for Mines and Co-operatives, Mr Zulu, Zambia Hansard, 10 August 1965.
18. Wage costs increased by 49 per cent between 1964/5 and 1966/7. Derived from M. Bostock, 'The Background to Participation', in M. Bostock and C. Harvey (eds.), Economic Independence and Zambian Copper: A Case Study in Foreign Investment (Praeger, New York, 1972), Table 5.3, p. 117.
19. Ibid.
20. Transport costs per ton of copper exported increased by some 34 per cent from 1964 to 1967. Zambia Mining Year Book, various issues.
21. Baldwin, Economic Development and Export Growth, pp. 34-5.

22. Martin, Minding Their Own Business, pp. 122-3.
23. For details of this change in pricing systems and the reasons for it, see O'Faircheallaigh, 'The Role of Foreign Investment in Mineral Development', pp. 169-70.
24. Republic of Zambia, Financial Report (Government Printer, Lusaka), various issues.
25. See the statement by the Minister for Mines and Cooperatives, Zambia Hansard, 10 August 1965; Republic of Zambia, First National Development Plan, 1966-1970 (Government Printer, Lusaka, 1966), p. 31.
26. The First National Development Plan, for example, repeatedly emphasised the need for increased production, but did not discuss the reasons for pursuing this aim or the requirements for its successful implementation. (See Chapter VI of the Plan, 'Mines'.)
27. Martin, Minding Their Own Business, p. 132.
28. What criticism there was came almost entirely from European members, whose motives were suspect in the eyes of many Zambians. See the statements by Burney (Zambia Hansard, 20 July 1967) and Stanley (Zambia Hansard, 4 August 1967), for example, and the reactions to them by Zambian members.
29. Martin, Minding Their Own Business, p. 133.
30. Ministerial responsibility for Mines changed hands five times between 1964 and 1969.
31. RST apparently deferred consideration of developments at Baluba and Chimbishi. Martin, Minding Their Own Business, p. 132.
32. For a discussion of the reasons for the Zambian government's dissatisfaction, see O'Faircheallaigh, 'The Role of Foreign Investment in Mineral Development', pp. 176-7.
33. Zambia's Economic Revolution, Address by His Excellency The President, Dr K.D. Kaunda, at Mulungushi, 19th April 1968 (Government Printer, Lusaka, n.d.), pp. 38-40, 43, 50.
34. 'Book value' normally reflects the historical cost of a company's assets less whatever has been provided for depreciation on them.
35. Zambia's Economic Revolution, pp. 49-50
36. A. Drysdall, 'Prospecting and Mining Activity, 1895-1970', in Bostock and Harvey, Economic Independence and Zambian Copper, p. 73.
37. These figures are derived from AAC, 'Statement by the Chairman, Mr. H.F. Oppenheimer, May 1968,' p.4.
38. Derived from Table 2.1.
39. Derived from BCL, Annual Reports, 1973-80.
40. Martin, Minding Their Own Business, p. 151.

41. Towards Complete Independence, Address by His Excellency the President, Dr K.D. Kaunda, to the UNIP National Council held at Matero Hall, Lusaka, 11 August 1969 (Government Printer, Lusaka, n.d.), pp.29-34.
42. Ibid., pp.36-7.
43. Ibid., p. 36.
44. Kaunda, Towards Complete Independence, pp.29-31, Zambia's Economic Revolution, pp.7, 38-40, 49-50; C. Harvey, 'Economic Independence', in Bostock and Harvey, Economic Independence and Zambian Copper, pp.3-22; M. Ndulo, 'The Requirement of Domestic Participation in New Mining Ventures in Zambia', African Social Research, 25 (June 1978), pp. 399-403, 412.
45. In certain cases this policy was based on economic as well as political grounds, as the government hoped it would encourage industrial growth in Zambia.
46. Copper Industry Service Bureau, Training for Zambianisation in the Mining Industry (Kitwe, n.d.), pp. 5, 12; information supplied by Mining Industry Manpower Services Unit (MIMSU).
47. Martin, Minding Their Own Business, p. 158.
48. Ibid., pp. 158-9.
49. The terms of the arrangement were contained in a Master Agreement and various subsidiary documents including management contracts, sales and marketing contracts, and the Memoranda and Articles of Association of the two new Governemnt mining companies. The contents of these documents are summarised in Bostock and Harvey, Economic Independence and Zambian Copper, Appendix A.
50. In RCM's case the minority shareholders were RST International (20 per cent), Zambia Copper Investments Limited (ZCI), a holding company for AAC's Zambian interests (12.5 per cent), and non RST shareholders in the former RST group mines (16.5 per cent). ZCI was the sole private shareholder in NCCM.
51. Bostock and Harvey, Economic Independence and Zambian Copper, p. 235.
52. Ibid.
53. These provisions were contained in the Income Tax (Amendment) Act 1970.
54. Figures for ZIMCO's liabilities were obtained from Bostock and Harvey, Economic Independence and Zambian Copper, p. 228; figures for dividend entitlements were derived from NCCM, Annual Reports, 1971-81 and RCM, Annual Reports, 1970-8.
55. RCM, Annual Reports, 1971 and 1977; NCCM, Annual Reports, 1971 and 1977; Zambia Mining Year Book,

1971, p. 43 and 1977, p. 31.
56. Bostock and Harvey have suggested a number of ways in which such a provision could have been formulated: Economic Independence and Zambian Copper, pp. 174-5.
57. Martin, Minding Their Own Business, p. 159.
58. Changes in production costs are relevant in this context because a high proportion of mining industry inputs are imported.
59. Zambia's balance of trade was in surplus by K384.1 million in 1973 and in deficit by K76 million in 1975. Republic of Zambia, Economic Report, 1976 and 1977.
60. As mentioned above, prior to 1970 the mining companies charged capital expenditures designed to maintain production against revenue. The Annual Reports of the individual companies did not give a breakdown of charges against revenue (see, for example, Nchanga Consolidated Copper Mines, Annual Report, 1966). Consequently it is impossible to gain an accurate picture of capital expenditures by individual mines.
61. In 1977, for example, production costs at the tailings leach plant were below K300 per tonne, while average costs for the NCCM mines were K752 per tonne. Interview with NCCM official, Chingola, 27 July 1978; NCCM, Annual Report, 1977, p. 27.
62. Derived from RST Limited, Annual Report, 1968, p.19, and RCM Annual Reports, 1971-4.
63. Derived from RCM, Annual Reports, 1971-4.
64. The Tailings Leach Plant at Chingola, for example, was initiated in 1970, while the period of major capital expenditures was 1973-5.
65. Republic of Zambia, Second National Development Plan, 1972-6 (Government Printer, Lusaka, 1971), p. 91.
66. For example with Noranda Mines Ltd. and Geomin.
67. R. Sklar, Corporate Power in an African State: The Political Impact of Multinational Mining Companies in Zambia (University of California Press, Berkeley, 1975), p. 54, n. 61.
68. All capital expenditure could be written off immediately against revenue in calculating profits, and losses could be carried forward indefinitely. These provisions were altered in 1973.
69. The Second National Development Plan states that mineral development is needed 'to increase employment opportunities, Government revenue, foreign exchange earnings, and development of rural areas' (p.85).
70. The higher the debt: equity ratio, the greater

the risk borne by the equity investor, and hence the greater his minimum expected rate of return. A return of 25 per cent on equity has been quoted as 'reasonable' for a project with a 1:1 debt : equity ratio, a low ratio by current standards. A. Gordon, 'The Prospects for New Mine Investment', in Bostock and Harvey, *Economic Independence and Zambian Copper*, p. 196.

71. This is illustrated by the case of Agip, for example, which has agreed to pay Zambia at least 10 per cent of the value of uranium production even if its operations turn out to be unprofitable. *Mining Journal*, 13 April 1979.

72. Interview with Director, Anglo American Central Africa Ltd., Lusaka, 1 August 1978.

73. Martin, *Minding Their Own Buiness*, p. 156.

74. Dr. Kaunda delivered his Matero speech on 11 August; agreement in principle on all the major issues was reached by 18 October.

75. Interview with Anthony Martin, Port Moresby, 8 June 1978.

76. Martin, *Minding Their Own Business*, p. 159.

77. It may or may not be coincidental that the total tax rate applied to mining companies in Chile in 1969 was also 73 per cent. Girvan, *Copper in Chile*, p. 30.

Chapter Five

NATIONAL VERSUS CORPORATE CONTROL OF MINERAL
RESOURCES IN ZAMBIA, 1964 - 1978

The Zambian government quickly came to realise the
weaknesses of the 1969 Take-over arrangements, and
in August 1973 announced their amendment. The
outstanding ZIMCO bonds would be redeemed
immediately, and the minority shareholders would
therefore lose their immunity from changes to
taxation and exchange control legislation. The
government took advantage of this situation to end
the immediate write-off of capital expenditure
allowed by the 1970 tax laws, and to impose a
withholding tax of 20 per cent on dividends payable
to foreign shareholders.
    Redemption of the bonds did not allow the
government to unilaterally alter the management and
other contracts or RCM's and NCCM's Articles of
Association. Negotiations were undertaken with AAC
and American Metal Climax (AMAX) (the principal
minority shareholder in RCM),[1] and it was agreed
that the management, sales and purchasing contracts
would be terminated. The government would establish
a state company to market Zambia's copper, while the
operating companies, RCM and NCCM, would become
self-managing. However AAC and AMAX would continue
to second senior technical and managerial personnel
and to supply, on request, certain services which
NCCM and RCM could not themselves provide. The
Articles of Association of RCM and NCCM were amended
so as to weaken the powers of the minority
shareholders. In particular, the requirement for
separate votes of 'A' and 'B' shareholders and
directors on certain specified issues was removed.
However, it was provided that a quorum which would
include two 'B' directors would have to be present
before the Boards of Directors could take decisions
on a (more limited) range of issues. These included
the raising of loans and participation in non-mining
activity, but excluded appropriations for capital or

126

exploration expenditure and amendment of the companies' Articles of Association.[2] Thus AAC and AMAX did not entirely lose their power of veto, since the 'B' directors could absent themselves and block certain initiatives, but its range was narrowed and the ability of the 'A' directors to further amend the Articles of Association would make its frequent use very unlikely. The 'B' directors also ceded the right to choose the Managing Directors of NCCM and RCM to the government.

The financial cost of these changes was high. In order to redeem the ZIMCO bonds, the government raised two Eurodollar loans totalling US$150 million, at interest rates well above those borne by the ZIMCO bonds. In addition, it paid K33 million to AAC and K22 million to AMAX in compensation for early termination of the various contracts.

Once again, the principal motivation for the government's initiative was the search for control. First, it was frustrated by the limitations on its power to amend taxation and exchange control legislation. The immediate write-off of capital expenditure meant that effective tax rates were low; for example, RCM paid only 19 per cent of its profits in taxation in 1972, NCCM 23 per cent in 1973,[3] as opposed to the nominal rate of 73 per cent. Tax liability was of course only being postponed, but the prospect of the mining companies paying low rates of tax was politically unpalatable, especially since prices and profits were on the increase after a number of lean years (1970-72).[4] In addition, the capital allowances had been granted to encourage expansion of copper production, but as yet there was little evidence that they were having the desired effect (see Table 4.1). (The low tax rates mentioned above resulted from write-off of capital expenditure designed to maintain production.) The whole question of investment policies was in fact still unsettled; in announcing the redemption of the ZIMCO bonds, President Kaunda asserted that the minority shareholders had used their powers to ensure that a high proportion of profits would continue to be distributed, and that loans would be used to finance what development did occur.[5]

The government was anxious to amend exchange control provisions so it could capture part of the substantial dividends then flowing to foreign investors. It expected to obtain an additional source of revenue by marketing Zambia's copper; it now felt competent to perform this function, and saw no reason why marketing fees (which amountd to about

K4.5 million in 1972/3)[6] should accrue to ACC and AMAX.

The government was also concerned with a number of non-financial matters. It had envisaged in 1970 that the management companies would gradually transfer managerial functions to RCM and NCCM, but this had not occurred to any great extent. In addition, it felt that the management companies had failed to amend their purchasing policies so as to reduce imports from South Africa. Finally, it was concerned with the slow pace of Zambianisation.[7]

This discussion raises once again the question of whether Zambia's interests were in fact in conflict with those of the foreign shareholders. In the following sections, the economic impact of copper mining is analysed, and this analysis should provide some relevant empirical evidence.

Generation of Government Revenue

As Table 5.1 illustrates, copper mining is an extremely important, though unstable, source of government revenue, accounting for 46 per cent of total revenue over the years 1964-75. This figure

Table 5.1: Government Revenue Generated by the Zambian Copper Industry, 1964-78 (K million)

| Year | Copper Revenue (A) | Total Revenue (B) | (A) as %of of (B) |
|------|--------------------|-------------------|-------------------|
| 1964 | 57 | 108 | 53 |
| 1965 | 134 | 189 | 71 |
| 1966 | 163 | 255 | 64 |
| 1967 | 146 | 276 | 53 |
| 1968 | 183 | 306 | 60 |
| 1969 | 237 | 401 | 59 |
| 1970 | 251 | 435 | 58 |
| 1971 | 114 | 316 | 36 |
| 1972 | 56 | 302 | 19 |
| 1973 | 108 | 385 | 29 |
| 1974 | 341 | 647 | 53 |
| 1975 | 59 | 448 | 13 |
| 1976 | 12 | 443 | 3 |
| 1977 | – | 498 | – |
| 1978 | – | 533 | – |

Source: Zambia Mining Year Book, various years.

does not fully reflect the copper industry's significance, since it excludes employee and indirect taxation[8] and revenue generated by ancillary and service industries dependent on copper mining.

Principally because of price fluctuations, copper revenues are extremely unstable, even over short time periods. For example, revenue was K57 million in 1964 and K163 million in 1966, K251 million in 1970 and K56 million in 1972. This instability causes serious economic problems. Since revenue fluctuations are unpredictable, government cannot plan appropriate levels of expenditure in advance.[9] And once a particular level of expenditure has been planned, major difficulties arise in adjusting it downwards or upwards to take account of the revenue which actually materialises. As a result, the government must often accumulate cash reserves in times of high prices and resort to deficit financing when revenues contract. If prices are depressed for an extended period of time, deficit financing imposes an unacceptable burden, and expenditure must be cut. Thus by 1977, for example, outstanding public debts had reached K950 million, with repayments and interest together equal to 28 per cent of total recurrent revenue in that year.[10] The government was compelled to reduce current and capital expenditure,[11] and in an economy where maintenance of employment and investment levels depends heavily on government expenditure, such action clearly has serious consequences.

In the longer term, perhaps the most serious effect of revenue fluctuations may be to undermine the attempt to deal with the underlying problem, Zambia's dependence on copper mining. That attempt must include steadily-expanding investment in agriculture and industry, but revenue fluctuations make it extremely difficult to sustain such investment.

Have foreign mining companies removed profits from Zambian jurisdiction, depriving the government of revenue? Since Zambia's copper is sold to independent fabricators, what is involved is the possibility that subsidiaries of AAC and/or RST (AMAX after 1970) operating outside Zambia over-charged the Zambian companies for mining equipment or other supplies. Lack of empirical data makes it difficult to deal with this issue, but one general point should be made. Both mining groups would have to be involved if over-charging was to escape detection; otherwise a simple comparison of invoices would reveal what was happening. As far as could be established, AAC alone has major shareholdings in non-Zambian companies which produce inputs for mining,[12] and there seems no reason why the RST group mines (or RCM after 1970) would transfer profits to AAC. Thus while the opportunity for transfer pricing existed in theory, it seems unlikely that it occurred.

### Generation of Employment and Wage Payments

Table 5.2 gives details of Zambian and expatriate employment in the copper industry over the period 1952-78. Mining employment is of considerable importance in national terms, accounting for 15 per cent of those in paid employment in 1975. Information regarding indirect employment is lacking, but the number employed in ancillary and service industries must be considerable.

A number of points emerge from the data in Table 5.2. Expatriate employment fell by 35 per cent in the four years after Independence, reflecting the rapid Zambianisation of semi-skilled and lower

Table 5.2: Employment in the Zambian Copper Industry, Selected Years 1952-70, and 1970-8

| Year | Expatriate | African | Total |
|------|-----------|---------|-------|
| 1952 | 5,504 | 36,668 | 42,172 |
| 1956 | 7,067 | 37,533 | 44,600 |
| 1960 | 7,528 | 36,806 | 44,334 |
| 1964 | 7,455 | 38,097 | 45,552 |
| 1968 | 4,845 | 43,198 | 48,043 |
| 1970 | 4,375 | 44,094 | 48,469 |
| 1971 | 4,751 | 44,997 | 49,748 |
| 1972 | 4,600 | 46,245 | 50,845 |
| 1973 | 4,505 | 48,287 | 52,792 |
| 1974 | 4,392 | 51,736 | 56,128 |
| 1975 | 4,495 | 52,992 | 57,487 |
| 1976 | 4,060 | 53,082 | 57,142 |
| 1977 | 3,609 | 55,446 | 59,055 |
| 1978 | 3,245 | 53,437 | 56,682 |

Source: Republic of Zambia, Report of the Commission of Inquiry into the Mining Industry (Government Printer, Lusaka, 1966), Appendix XVI; Zambia Mining Year Book, various issues.

supervisory positions. It changed little over the period 1968-73, because Zambianisation proceeded much more slowly at higher skill levels. Thereafter, it fell substantially, by 30 per cent over the period 1973-8. This decline was not entirely due to Zambianisation; the mining companies were unable to find replacements for some of the expatriates who departed.[13] Shortage of skilled expatriate staff has important consequences for the efficiency of mining operations, both immediately and in the longer term. It has contributed to the difficulties encountered in maintaining copper production in recent years,[14] and its impact is likely to increase as time goes on (see pp. 153-4 below).

Various factors account for the high turnover among expatriates[15] and the inability of the companies to find replacements. Particularly important is the lack of security offered by fixed-term contract employment; other factors are the politically unstable situation in Southern Africa, shortages of foodstuffs and other items in Zambia, urban violence, and delays in obtaining clearance for repatriation of savings and severance payments.[16]

Table 5.2 also shows that Zambian employment grew by 11,300 between 1970 and 1977. The reasons for this increase are not immediately apparent. Zambianisation had some impact, but over the period 1973-7, for example, Zambian employment in those job categories subject to Zambianisation increased by only 1,136,[17] while total Zambian employment grew by 7,159. Production did not increase over the period 1970-7 (see Table 4.1), while development work was at a substantially lower level in 1977 than in 1970.[18] Only two explanations for the increase in Zambian employment present themselves. A deliberate policy of increasing such employment may have been pursued in the wake of the 1969 Take-over, either by adopting more labour-intensive mining techniques or simply by 'feather-bedding'. Government officials deny that this has occurred,[19] but the only alternative explanation is that productivity has fallen, requiring expansion of the workforce to maintain production. Productivity would have to decline very substantially to necessitate such a large increase in employment, and there is no conclusive evidence that this has occurred.[20]

Generation of employment on the Copperbelt has carried an opportunity cost in that labour has been withdrawn from subsistence farming, and in areas close to the mines agricultural activity has suffered severely (see p. 143 below). Those who migrated to the Copperbelt were of course given an opportunity to substantially increase their incomes.

As mentioned in the previous Chapter, employment policies were identified as another area in which corporate and national interests might clash. Zambia's mining industry could not have competed on the world market had it not been based on large-scale productive units using substantial amounts of capital and applying a high level of technological expertise,[21] and this placed a major constraint on the industry's potential for creating employment. Nevertheless, as R.E. Baldwin has demonstrated,[22] the mining companies did significantly modify their mining methods in line with shifts in the relative costs of semi-skilled labour, skilled labour, and capital, and these changes have exercised a significant impact on total employment. In the late 1940s and the early 1950s, for example, average African wages more than doubled, with the result that the cost of semi-skilled labour grew much more quickly than that of skilled labour and equipment. The companies responded by initiating a far-reaching mechanisation programme which allowed them to double production over the period 1952-60 without increasing their use of African (that is semi-skilled) labour (see Tables 4.1 and 5.2).

Were corporate employment policies in conflict with national interests? From the point of view of Zambia's resource endowment, it might have been appropriate to maximise use of relatively plentiful semi-skilled labour and minimise use of scarce capital and skilled labour. However, this would not be so if additional government revenue could be generated by pursuing profit-maximising factor combinations and that revenue expended so as to create more employment than would result from pursuit of the (limited) opportunities for substitution of labour for capital in the mines. A judgement on the issue depends, therefore, on whether the mining industry was expected to contribute towards economic development by maximising direct employment or by providing revenues to fund government expenditures. As mentioned above, the Zambian government failed to come to grips with this issue, and it is therefore impossible to decide whether a conflict of interest existed.

Table 5.3 gives details of wage and salary payments in selected years over the period 1952-77.

Table 5.4 compares average expatriate with average
African incomes in the same years, and illustrates
the huge income differentials which existed in the
early and mid 1950s. These gradually narrowed as a
result of pay increases won by African mineworkers
and of the large-scale promotion of Zambians to more
skilled positions after Independence. However, by
1977 the differential was still very large (8:1 in
favour of expatriates).

Table 5.3: Wage and Salary Payments by the Zambian
Copper Mining Companies, Selected Years 1952-73, and
1973-7 (K000s)*

| Year | Expatriates | Africans | Total |
|------|-------------|----------|-------|
| 1952 | 16,512 | 6,307 | 22,819 |
| 1956 | 32,438 | 12,461 | 44,899 |
| 1960 | 32,521 | 18,992 | 51,513 |
| 1965 | 42,176 | 28,344 | 70,520 |
| 1968 | 36,822 | 56,157 | 92,979 |
| 1973 | 45,460 | 81,075 | 126,535 |
| 1974 | 51,533 | 93,309 | 144,842 |
| 1975 | 52,970 | 95,404 | 148,374 |
| 1976 | 54,409 | 12,412 | 166,821 |
| 1977 | 61,866 | 125,823 | 187,689 |

* Figures for years prior to 1968 have been coverted
to Kwacha at the rate of £sterling = 2 Kwacha.
Source: Baldwin, Economic Development and Export
Growth, Table 4.4, p. 90; Republic of Zambia,
Commission of Inquiry, Appendix XVI; information
supplied by MIMSU.

Creation of Additional Economic Activity through
Linkage Development

Forward Linkage. Almost all of Zambia's copper
production is integrated to the refining stage.
Copper fabricating industries have not been
established, with the exception of a small plant
which produces copper wire and cable for the
domestic market and absorbs less than 0.5 per cent
of total copper output. Copper fabrication could
create important benefits for Zambia, particularly
by generating additional employment and higher and
more stable foreign exchange receipts.[23] However,
economic constraints severely limit Zambia's

Table 5.4: Average Expatriate and African Incomes on the Zambian Copper Mines, Selected Years 1952-73, and 1973-7 (Kwacha)

| Year | Average Expatriate (A) | Average African (B) | Ratio of (A) to (B) |
|------|------------------------|---------------------|---------------------|
| 1952 | 3,000 | 172 | 17 : 1 |
| 1956 | 4,590 | 332 | 14 : 1 |
| 1960 | 4,320 | 516 | 8 : 1 |
| 1965 | 5,871 | 716 | 8 : 1 |
| 1968 | 6,847 | 1,300 | 5 : 1 |
| 1973 | 10,091 | 1,679 | 6 : 1 |
| 1974 | 11,773 | 1,804 | 7 : 1 |
| 1975 | 11,784 | 1,803 | 7 : 1 |
| 1976 | 13,401 | 2,117 | 6 : 1 |
| 1977 | 17,142 | 2,269 | 8 : 1 |

Source: Table 5.2 and 5.3.

potential in this regard, the most significant of which arise from Zambia's distance from the principal consuming areas, Europe, Japan and North America, the tariffs which developed countries impose on fabricated imports, and the high degree of skilled labour involved in what are frequently technologically-complex processes.

During the mid-1960s, the Zambian government was anxious to establish a plant which would supply copper wire and cable and semi-manufactured products to the domestic and East African markets. AAC and RST were reluctant to become involved in this venture, partly because of their reluctance to undertake any new commitment for the time being, partly because of the economic constraints mentioned above.[24] The government decided to proceed alone, and formed a consortium headed by a major US copper producer and fabricator, Phelps Dodge, to examine the project. Faced with the prospect that potentially rival companies might establish a fabricating plant in Zambia, AAC and RST modified their attitudes and joined in establishing the copper wire and cable plant mentioned above.

Backward Linkage. Prior to 1964, the mining industry's domestic expenditures were dominated by payments for electricity, transport services and construction.[25] Few manufactured inputs were produced locally, reflecting the competitive advantage derived by Rhodesian and South African producers from economies of scale associated with the larger size of their domestic markets and, after 1953, the discriminatory policies of the Federal government in Salisbury. After Independence the Zambian government offered protection and encouragement to new manufacturing concerns supplying mining inputs. Items now produced locally include chemicals, basic metal products, machinery, explosives, copper wire and cable, and sulphuric acid.

Some of these industries have not achieved minimum economies of scale,[26] with the result that the mining companies frequently pay well in excess of import parity prices for the goods involved. This raises once again the question of precisely what role the copper industry is expected to play in Zambia's economy. In effect, a portion of its sales receipts is being used to subsidise inefficient industries, at the expense of profits and so government revenue. That revenue might make a greater contribution to economic development if it were collected and directed into other sectors of the economy. On the other hand, some of the industries involved may create external economies which makes the support they receive worthwhile. Obviously a carefully selective approach is required, but there is little evidence that such an approach was applied in the years after Independence.[27]

With only one exception conflicts of interests have apparently not arisen between the mining companies and the government regarding use of domestic inputs.[28] The exception concerned the substitution of domestic coal for imported material in 1965/6. Zambian coal cost substantially more than imports from the Wankie colliery in Rhodesia, and AAC had an additional interest in importing coal since it owned the Wankie colliery. However, both responded immediately to Zambian government directives to reduce imports, and assisted in the development of domestic coal reserves.[29]

Economic constraints impose major limitations on Zambia's capacity to develop forward and backward linkages. However, considerable progress has been made in those areas where opportunities have been available. In pursuing these opportunities, the Zambian government found its interests in conflict with those of the mining companies on two occasions, but on both it was able to achieve its objectives, in one case by involving other corporate interests,

Control of Mineral Resources in Zambia

in the second by exercising its governmental powers
in a routine manner.

Net Foreign Exchange Inflow

Copper exports provide about 95 per cent of Zambia's
export revenue, but as Table 5.5 illustrates they
represent a very unstable source of foreign
exchange. Prolonged periods of low copper prices
cause serious foreign currency shortages,
restricting the availability of essential imports of
goods and services. This problem became so acute
during 1977 that in some cases allocations of
foreign exchange to importers were reduced to 10 -
25 per cent of normal, and severe shortages of many
imports resulted.

Table 5.5: Zambia's Copper Exports and Total Exports
by Value, 1964-78 (K million)

| Year | Copper Exports* | Total Exports | Copper as % of total |
|------|------|------|------|
| 1964 | 302 | 327 | 92 |
| 1965 | 348 | 375 | 93 |
| 1966 | 466 | 490 | 95 |
| 1967 | 440 | 467 | 94 |
| 1968 | 519 | 538 | 96 |
| 1969 | 729 | 752 | 97 |
| 1970 | 687 | 710 | 97 |
| 1971 | 454 | 480 | 95 |
| 1972 | 499 | 536 | 93 |
| 1973 | 703 | 738 | 95 |
| 1974 | 847 | 900 | 94 |
| 1975 | 479 | 518 | 93 |
| 1976 | 705 | 749 | 94 |
| 1977 | 661 | 706 | 94 |
| 1978 | 608 | 649 | 94 |

* Includes exports of cobalt which is mined as a by-
product of copper.
Source: Zambia Mining Year Book, various issues.

Due to lack of relevant statistical data, it is
not possible to calculate the net foreign exchange
inflow generated by Zambia's copper industry on a
basis similar to that applied in the previous case
study. However, it is possible to estimate the net
effect of NCCM's exports on current account for a

year of high prices (1973/4) and one of low prices (1975/6). In 1973/4, NCCM's exports were valued at K555 million. Non-wage production costs and capital expenditure totalled K241 million,[30] of which NCCM's Chief Economist estimates that 50 per cent or K120.5 million were expended abroad.[31] The import content of domestic purchases by the mining companies has been estimated at 28 per cent,[32] indicating induced imports by NCCM of K33.5 million. Dividend payments to foreign shareholders were K32.8 million,[33] while expatriate remittances are estimated at K8.8 million.[34] Interest payments abroad were about K2 million.[35] Thus it is estimated that foreign exchange outflows totalled about K198 million, implying a net positive effect of K357 million, equal to 64 per cent of export receipts. Government revenue accounted for K198 million or 56 per cent of this sum;[36] wages and salaries (20 per cent) and domestic purchases of goods and services (24 per cent) accounted for the remainder.[37]

In 1975/6 NCCM's exports were K327 million, and total foreign exchange outflows are estimated at K202.7 million,[38] implying a net foreign exchange inflow equal to only 38 per cent of export receipts. No government revenue was generated in 1975/6.

The difference between the net foreign exchange inflow estimates for the two years (64 versus 38 per cent) raises an important point. When copper prices fall, not only do export receipts decline, but the proportion of those receipts accruing as net foreign exchange inflow also falls. This is because government revenue (which is itself sensitive to declining prices) accounts for such a high proportion of the net inflow, and it partly explains the severity with which declining copper prices affect Zambia's capacity to import goods and services.

Provision of Scarce Capital Resources

Foreign equity capital played an indispensable role in the initial development of Zambia's copper mines, as investment on the scale required could not have been generated internally. Once mining operations were well established, equity investment became less important, as the bulk of capital expenditure could be financed from retained profits. Foreign capital still plays a major role in Zambia's mining industry, though the type of investment involved has changed. Since 1970 there has been a very large increase in the use of loan capital, a form of finance almost unknown in the industry until 1968/9, when the AAC group mines raised two major loans from Japanese companies. Table 5.6 gives a breakdown of RCM's and NCCM's capital expenditures for the period 1971-7, according to the manner in which they were financed (loans versus reinvested profits). This shows that 53 per cent of RCM's expenditures and 38 per cent of NCCM's were financed by loan capital. A substantial proportion of this capital has come from outside Zambia: in 1977, 60 per cent of RCM's outstanding loans and 76 per cent of NCCM's had been raised abroad.[39] These figures underestimate the role of foreign capital, as some 'local' loans are refinanced with foreign banks.

No foreign equity was invested in mineral development (other than in exploration) between 1969 and 1978. The minority shareholders in RCM and NCCM refused to participate in share issues[40], while no other companies undertook major projects. This picture will almost certainly have to change if major new mineral developments are to be undertaken. These would require substantial equity investment, and it is doubtful whether Zambia could supply that investment from internal sources without depriving other sectors of much-needed capital; the mining industry already absorbs about 20 per cent of total public and private capital investment.[41]

The method in which capital investment was being financed over the period 1971-3 was one reason given by the Zambian government for its redemption of the ZIMCO bonds. However, increased use of loan capital was not in itself the problem. When negotiating the Take-over terms, the government had placed considerable value on the fact that the mining companies carried little or no debt burden and could therefore raise substantial loans to fund development programmes.[42] Raising of loans required the approval of 'A' directors, and the government could have prevented use of loan capital if it had so desired. The real problem was that a higher share of profits was not being reinvested in addition to the loans. As mentioned above, the minority shareholders succeeded in keeping reinvestment rates at a low level after 1970, and their interests were clearly in conflict with the government's in this instance.

Provision of Technical and Other Skills

Until the late 1950s, the mining companies obtained their skilled personnel almost entirely by

Table 5.6: Funding of Capital Expenditure in the Zambian Copper Industry, 1971-7 (K million)

| Year | Capital Expenditure | Appropriated from Profits | | Loan Finance | |
|---|---|---|---|---|---|
| | | Amount | % | Amount | % |
| NCCM | | | | | |
| 1971* | 43 | 43 | 100 | - | - |
| 1972 | 42 | 30 | 71 | 12 | 29 |
| 1973 | 59 | 39 | 66 | 20 | 34 |
| 1974 | 69 | 46 | 67 | 23 | 33 |
| 1975 | 59 | 40 | 68 | 19 | 32 |
| 1976 | 39 | 5 | 13 | 34 | 87 |
| 1977 | 16 | - | - | 16 | 100 |
| Total | 327 | 203 | 62 | 124 | 38 |
| RCM | | | | | |
| 1971 | 28 | 28 | 100 | - | - |
| 1972 | 34 | 12 | 35 | 22 | 65 |
| 1973 | 28 | 11 | 39 | 17 | 61 |
| 1974 | 27 | 20 | 74 | 7 | 26 |
| 1975 | 37 | 24 | 65 | 13 | 35 |
| 1976 | 34 | - | - | 34 | 100 |
| 1977 | 56 | 20 | 36 | 36 | 64 |
| Total | 244 | 115 | 47 | 129 | 53 |

* Includes expenditures made in 1970.
Source: NCCM, Annual Reports, 1971-7; RCM, Annual Reports, 1971-7.

recruiting qualified expatriates and training
(exclusively white) apprentices; Africans were
taught only the most basic industrial skills. Since
the late 1940s, AAC and RST had been anxious to
train Africans to replace expensive expatriate
workers, but the opposition of the European union
effectively precluded on-the-job training, while
Northern Rhodesia's schools failed to produce
suitable Zambian candidates for formal advanced
education. In the early 1960s the companies
initiated a broad training programme, eventually
operating at three levels. The first was designed to
equip Zambians for semi-skilled and lower
supervisory positions predominantly by on-the-job
training, the second to produce Zambian artisans and
technicians by a mixture of formal trade school
education and on-the-job training, the third to help
develop a supply of technical and professional
graduates, principally through a programme of
scholarships and sponsorships.

Progress on the first level was rapid, and by
1968 the programme was essentially complete. Two
major problems have arisen regarding the second,
both involving the role of expatriates in on-the-job
training. First, expatriate skilled labour has been
in short supply for much of the period since 1964;
the available staff have been fully occupied in
maintaining copper production, and have had little
time to assist apprentices assigned to them. Second,
expatriate workers are aware that they are 'training
themselves out of a job', and some are reluctant to
pass on their skills, particularly if their own job
mobility is limited.[43] The outcome is that newly-
qualified Zambian staff assigned to the operating
divisions are frequently unable to deal with
practical work situations,[44] and have either to be
returned for further training or left to acquire
skills on a trial-and-error basis. The first must
cause frustration and resentment, the second
inefficiency and, where underground work is
involved, danger.

Training of technical and professional
graduates has progressed very slowly, for two main
reasons. First, it was some time before appropriate
educational institutions were established.[45] Second,
there has been a shortage of suitable candidates for
appropriate degree courses. The mining industry's
main requirements are in the engineering and related
fields, and third level institutions consequently
need a high intake of school leavers with

backgrounds in mathematics and science. Unfortunately, it is precisely in this area that school output is weakest; in 1975, for example, only 30 'O' level candidates examined in mathematics, physics and chemistry received the grades required for admission to university level in each subject.[46]

As a result of these factors, the number of Zambian graduates employed in the mining industry has remained very low: in 1978 only 92 were working in mining/geology, metallurgy/chemistry and engineering.[47] This picture will improve as additional students graduate from the University of Zambia, but if the current rate of graduate output remains unchanged, it will be about twenty years before all graduate positions can be localised, assuming no wastage of Zambian staff.[48]

Training in the mining industry is now the responsibility of the Zambian government, but the foreign investors (especially AAC) still play a crucial role in the provision of skills at the highest technical and professional level. Staff with the ability and experience to fill positions at this level are in short supply (both in Central and Southern Africa and world-wide), and many would not choose to work in Zambia without the security of employment and opportunities for advancement which they derive from being attached to a major international mining concern. Thus a number of AAC's senior staff in Zambia refused to transfer to NCCM, anad are now seconded to NCCM by AAC.[49]

The Zambian government has cited the slow pace of Zambianisation as one reason for seeking increased control over the copper industry. However, it is not simply a question of the government being wholly committed to rapid Zambianisation and the companies or later the minority shareholders being against it. In general terms, both parties have feared that rapid Zambianisation would reduce efficiency and so profitability. Emphasis has been placed on 'proper and efficient' Zambianisation, not just by company officials, but also by the Department of Mines,[50] and this may have led to over-caution in some cases. On the other hand, both parties have had a strong financial incentive to replace expatriate staff who inflate wage bills and so reduce profits.

In more specific terms, the principal constraint on Zambianisation at the technical and professional level has arisen from the nature of the educational system, and what has been most lacking in this area is a government initiative to expand the supply of suitably-qualified candidates. The government's desire to obtain high school and university graduates to fill its own manpower requirements has probably contributed to the development of an educational system with a low output of science-oriented students, and to the absence of serious attempts to rectify the situation. The mining companies have not actively retarded progress in this area. They have not denied financial assistance to available candidates for third level education, and have apparently not discriminated against Zambian graduate employees.[51] However, they could have adopted a more positive attitude, for example by attempting to isolate routine components in the work of expatriate employees (for instance chemists) which could be performed by Zambian technicians, an approach which is now being adopted with considerable success.[52]

Turning to the skilled manual level, responsibility for the scarcity of expatriate staff to provide on-the-job training is shared by the companies and the government. The companies have not (until recently) employed expatriates whose sole or major responsibility would be training, probably because of the expense involved.[53] This option could usefully have been examined at a much earlier date. On the other hand, the government could have done more to encourage expatriate recruitment, for example by ensuring rapid exchange control clearance for remittances.

The attitude of expatriate workers to Zambian apprentices and trainees raises more complex questions. Management can do little when faced with employees who refuse to pass on their skills. Any attempt to keep a rigid check on trainees and to penalise employees where lack of progress was evident would almost certainly be counter-productive; it would be resented by expatriates, and would probably lead to further resignations and compound the problem of finding replacements. The difficulty in this area is a deep-seated one. Because of their key role in the industry, management is very loath to take action against expatriates who are retarding Zambianisation; the risk in terms of loss of efficiency is simply too high. No new initiatives have been taken in this area since 1973, indicating that this factor has acted as a constraint on the Zambian government, just as it did on the foreign investors.

The requirement for rapid Zambianisation has conflicted with other interests being pursued by the government or the companies or both, and this explains why progress has frequently been slow. Corporate and national interests may have come into conflict on this issue, but only in the sense that government would have liked the companies to be more enterprising in those areas where its other interests were not threatened.

Provision of Infrastructure

As mentioned earlier, the spin-off effects of mine infrastructure have been limited by the Copperbelt's geographical concentration. The mines are served by only one major railway line and one bitumenised road, both of which follow the same route. This situation may change if Zambia's widely-dispersed uranium deposits or other orebodies away from the line-of-rail are developed.

Absorption of Domestic Resources

Foreign mining companies have not mobilised significant amounts of domestic capital in Zambia, but their activities have absorbed substantial amounts of agricultural labour over an extended period of time. Anne Seidman has estimated that some villages have lost up to 60 per cent of their male populations in the 20-40 age group, leading to over-use of existing cultivated land and a serious decline in productivity.[54] Mining has also absorbed some agricultural land, but because of the Copperbelt's geographical concentration its impact in this regard has not been widely felt.

Impact on Consumption Patterns

The mining industry has had a major impact on consumption patterns in Zambia. It has drawn into close contact a large expatriate community and a substantial proportion of Zambia's population, with the result that Zambians have attempted to emulate expatriate patterns of consumption. This has had four major effects. First, it has fuelled Zambian demands for incomes on a par with those of expatriates, a point dealt with below in discussing income distribution. Second, it has created a substantial demand for imported consumer goods and services which, especially in times of low copper prices, absorbs foreign exchange needed to finance imports of essential goods and services. Third, Zambians frequently (indeed usually) fail to achieve the income levels required to support expatriate standards of consumption, and this sometimes leads to corruption or malpractice in the civil service and government. The Reports of Zambia's Auditor-General list many cases of misappropriation of funds by public servants,[55] while at least two serious cases of financial malpractice at the political level have been revealed.[56] Finally, there has been a spill-over effect into the countryside; rural Zambians have become aware of the consumption levels enjoyed by many urban dwellers, and this has contributed to rural-urban migration with its attendant social and economic problems.[57]

It is difficult to envisage how Zambia could have obtained the benefits of mineral development without incurring some costs as a result of changed consumption patterns, but these costs would probably have been considerably less had Zambianisation not been delayed for so long.

Wage Levels and Income Distribution

The mining industry has had a major impact on income distribution, for two reasons. First, because the industry accounts for such a large proportion of the wealth generated in Zambia, changes in income distribution within the mining sector have a major impact on national income distribution. Second, the mining industry's payments to labour influence labour payments in other sectors of the economy through a 'demonstration effect', and this in turn has a significant impact on national income distribution. In both cases, the mining industry's wage policies are a crucial factor.

During the period 1964-6, Zambian mineworkers received a number of substantial wage increases, through collective bargaining with the mining companies (12 per cent in 1964, 11 per cent in January 1966), and through a government award (22 per cent in August 1966). These wage increases resulted in a major redistribution of incomes from mining companies to Zambian labour[58]. In one sense, it was clearly preferable from Zambia's point of view that income should flow to nationals rather than to foreign companies who would probably remit a large part of it. However, it must be remembered that some 60-70 per cent of gross profits were absorbed by taxation, and thus redistribution of profits in favour of labour would also involve redistribution of government revenue in favour of labour. Any such development would have important

144

implications for domestic income distribution, as we
shall see below.

Wage increases in the mining industry had a
major impact in other sectors of the economy. The
industry acted as a 'wage leader',[59] with other
workers attempting to emulate the miners, and pressure was exerted on the government (as an employer
and as an industrial arbitrator) and on private
companies to grant substantial pay increases. This
trend was particulary evident after the government-
appointed Brown Commission awarded miners a 22 per
cent pay rise in August 1966. Within a year, public
service workers had obtained increases of 20 to 80
per cent, construction workers 33 per cent, agricultural workers 30 per cent, and workers in service
industries 25 to 55 per cent. The search for higher
wages cannot be solely attributed to the operation
of a 'demonstration effect'. Independence raised the
expectations of all Zambian workers, and the high
level of expenditure permitted by buoyant copper
prices increased demand for skilled and semi-skilled
labour. Nevertheless, events in the mining industry
did provide a catalyst. Even more importantly, wages
in the industry were already considerably higher
than in other sectors, and large percentage
increases in miners' wages meant that the level of
expectations of other workers (and consequently
their wage demands) would be high.

General wage increases had a number of
important economic effects, with one of the most
significant resulting from their impact on
government finances. In the mid 1960s, the
government employed about 65,000 people or 30 per
cent of the wage-earning workforce, and wage rises
consequently had a major impact on current
expenditure. Pay increases in the mining industry
also had a direct effect on government finances
since they reduced profits and so taxation.[60] The
impact of these developments was not very evident
during the mid-1960s, as high copper prices ensured
that revenue exceeded all capital and current
expenditure. However, high wage levels had been
established in the public service and the mining
industry, and these did not fall when copper prices
declined, with the result that sufficient revenue
was not available to maintain capital expenditure.[61]
The point has already been made that one of Zambia's
principal goals, economic diversification, will not
be achieved unless a substantial level of capital
investment can be maintained.

Wage increases also had a significant though

unquantifiable effect on employment. Labour costs represent a substantial proportion of total production costs in the manufacturing and agricultural sectors,[62] and the large pay increases of 1964-9 had a major impact on their employment-generating potential, for two reasons. First, they encouraged use of capital-intensive techniques. Second, they reduced Zambia's competitiveness as a potential exporter (especially in the East African region where wage rates were already lower than in Zambia), and increased the competitiveness of imports. Thus opportunities for industrial and agricultural development based on export markets and import substitution were reduced.[63] Partly for these reasons, growth in employment over the period 1964-70 was not as rapid as had been anticipated - 83,000 as against the minimum expected figure of 100,000.[64] In the late 1960s and early 1970s, the growth in employment was not maintained. At the same time, the urban population grew at about 8 per cent per annum, with the result that the number of urban unemployed increased by an estimated 8,000 to 10,000 a year.[65]

It was in fact unlikely that expansion in industrial employment could have matched the growth in urban population, regardless of wage rates.[66] Since much of that growth resulted from rural-urban migration, any attempt at dealing with unemployment would have to focus on rural development; thus the impact of wage increases on the prospects for such development was especially significant.

Pay increases for wage-earners had a major impact on distribution of income between rural and urban Zambians. The income of peasant farmers did not increase at nearly as high a rate as did urban wages. Over the years 1964-8, average peasant earnings increased by 53 per cent in money terms; average earnings of wage-earners outside the mines increased by 125 per cent, those of mineworkers by 105 per cent. In addition, urban wage increases affected the terms of trade between rural and urban areas. Goods and services provided by the urban to the rural sector became more expensive while the price of agricultural goods rose much more slowly; in real terms, Zambian rural incomes increased by a total of only 4 per cent over the period 1964-8, while those of Zambian wage-earners outside the mines increased by 52 per cent and those of Zambian mineworkers by 35 per cent.[67] Thus the additional wealth accruing to Zambians over this period[68] was very unevenly distributed, further skewing income distribution in favour of the urban areas. In 1964, wage earners outside the mines earned, on average, three times as much as peasant farmers, mine workers seven times as much (at 1968 prices); in 1968, the equivalent figures were four and a half times and nine times.[69]

When the gap between rural and urban incomes expands, rural-urban migration increases. This in turn weakens the prospects for the rural development which might stem further migration as it creates labour shortages, frequently deprives villages of their most enterprising inhabitants, and can make it uneconomic to maintain existing services or provide additional ones. At the same time, wage increases reduce the government's capacity to increase rural incomes, as less money is available to finance services and the capital investment which might create income-generating assets.

It is difficult to see how the Zambian government could have avoided these problems. It inherited a mining industry in which huge wage differentials separated blacks and whites, and there was very little it could do to prevent black workers from attempting to change this situation. Having conceded the mineworkers' demands, it would have found it very difficult to resist those of other workers, or to avoid the consequences of not doing so. However, its inability to avoid these difficulties had important consequences for its mineral policies, consequences which were not fully taken into account.

First, the government's desire for rural development called for a redistribution of income in favour of rural Zambians. From this perspective, it would have been preferable to allow the mining companies to pursue profit-maximising factor combinations, thus maximising tax revenues and the government's ability to divert resources into the rural sector. However, there are some indications that this has not been done since the 1969 Take-over (see above); certainly government statements on employment policies show little awareness of their implications for income distribution.[70] Similarly, the policy of encouraging inefficient production of mining inputs effectively redistributed income from the government to urban workers.

Second, it was apparent that the continued presence of a large number of expatriate miners would continue to exercise an upward effect on Zambian wage levels. It was essential that Zambianisation proceed as rapidly as possible; yet, as mentioned above, the government displayed little urgency in this area. Once again, the implications of policy alternatives had not been fully spelt out.

Ownership and Control of Mineral Resources

Before proceeding further, two other areas of potential conflict should be mentioned. The first involved the issue of the mining companies' purchasing policies. From what has already been said on this subject, it is apparent that a real conflict of interest did exist in this area: the companies were anxious to maintain their links with Southern Africa, the government wished to break them.

The second, which concerns mining practices, is more complex because the issues involved are technical in nature and because disagreements frequently occur even among disinterested experts as to what constitutes 'best mining practice'. The government believed that the companies might adopt practices inimical to Zambia's interests, for example by high-grading or by allowing excessive dilution of ore. High-grading can be a perfectly sound mining practice in certain circumstances; what the government feared was that the companies would continually extract the richest ores to maximise profits in the short term, leaving behind ore which could have been extracted profitably along with the richer material but which would be uneconomic to mine on its own. A certain degree of dilution is inevitable in any mining operation; however, excessive dilution might occur where companies were unwilling to invest in facilities for disposal of waste prior to processing, leading to higher operating costs and lower profits. As regards both these practices, the basic issue was the discount rate applied by the companies to future earnings; application of a high rate might make it worth their while to sacrifice lower-grade ore (and thus reduce total profits) in order to maximise short-term profits, and not worth their while to invest in waste disposal facilities so as to reduce future operating costs and increase future profits.

As mentioned already, it is difficult to obtain conclusive evidence on questions of this type. However, at least one independent authority believes that the companies did engage in systematic high-grading, and that this resulted in loss of workable ore,[71] while it is certainly true that their time horizons were short in the period after Independence. Thus the government's belief that conflicts of interest might exist in this area was probably justified.

In sum, five areas have been identified where conflicts of interest did exist or may have existed: pricing of imported goods, sourcing of imports, mining practices, Zambianisation, and investment policies. Was majority state ownership of the operating companies the best way to ensure that national interests would be protected in these areas? One point is immediately clear - majority ownership would not in itself ensure such an outcome. The government's directors and their advisory staff would have to be sufficiently qualified and well-informed to defend Zambia's interests in a situation where ownership and control would be shared with corporations who were extremely knowledgeable about the Zambian and international mining industries.

Information was of crucial importance as regards pricing and sourcing of imports. Import pricing could not be monitored unless information was available regarding prices offered by suppliers other than those affiliated to the parent companies. If imports from Southern Africa were to be reduced, it would be necessary to know which items could be obtained elsewhere and where to obtain them.[72] Armed with the relevant information, the government's directors could press its case; without it, they could do little. Monitoring of mining practices would require a degree of technical expertise and thus the recruitment of at least a small number of specialised staff who could carry out selective checks on working methods on a random basis.

The issue of investment policies is more complex. Its most obvious aspect was the unwillingness of the foreign investors to reinvest profits. In the absence of taxation measures which might have persuaded them to change their minds, the only alternative was for the government to obtain full control over dividend policy. To do this, it apparently had to obtain majority ownership; it was obvious that exchange control over dividend remittances, for example, had no effect. But there was a second dimension to this issue. Any investment decision depends on how the technical and economic feasibility of the prospective project is assessed. For example, if recovery of metal from waste dumps was being considered, it would be necessary to determine whether or to what extent it was physically possible to do this. The conclusions would in turn influence assessments of economic viability, which would also be based on projections of future metal prices and production costs. If these functions were carried out by the minority shareholders, government directors would be powerless to question their conclusions unless they could call on independent experts who could examine the methods and assumptions used in determining technical and economic viability. Government control over dividend policies would have little impact if the minority shareholders continued to decide what was and was not worth spending money on.

To understand the importance of this point, it is necessary to outline briefly the structure of decision-making in the mining industry. Since the 1969 Take-over, the position in theory is that the government sets broad policy guidelines for the industry; these are conveyed to the company boards, which formulate specific policies accordingly. The Managing Directors exercise executive power and translate these policies into action.[73] This may be the way in which decision-making actually occurs on issues such as wages policy, but in the crucial area of investment policies the situation is in fact more complex.

Initiatives which lead to investment decisions frequently come from operating divisions, where staff see a need or an opportunity for investment. The initial assessment work is carried out by technical staff at this level, in co-operation with accounting and finance staff from the Centralised Services Divison. If it is felt that the project warrants further expenditure, it is submitted to the Technical Director.[74] This individual plays a crucial role in decision-making. He views competing requests for funds and decides which, if any, will be forwarded to the Managing Director and the Board for their consideration.[75] If the Board initiates proposals itself, they must be submitted to the Technical Director and his staff, who assess their technical and economic viability. Thus if the government directors were not simply to rubber-stamp investment proposals deemed acceptable by the minority shareholders, they would require access to an independent source of technical expertise. The government would not have to duplicate every feasibility study, but only to ensure that its directors could request a second opinion if their investment proposals were declared unviable, and that projects suggested by the operating divisions and declared unviable by the technical staff could be re-examined.

As far as is known, an information-gathering

system was not established after 1970, which may
have contributed to the failure of the Take-over to
give the government effective control, though other
factors were of course more important. The
government did show awareness of the need for
technical expertise, and MINDECO was given the task
of monitoring the management and operating
companies. It recruited a small but high-qualified
staff which included at least one specialist in each
of the main areas of mining activity, for example
geology, metallurgy and engineering. MINDECO was
soon operating as an effective monitoring agency,
and undertaking detailed technical studies regarding
various aspects of mineral development.[76]
    The legal and other restrictions contained in
the 1969 Take-over arrangements prevented the
government from gaining effective control over the
copper industry, despite the expertise it had
developed, and it reacted by terminating the
management contracts. The government subsequently
took the view that since the operating companies
were self-managing there was no need to monitor
their activities.[77] MINDECO's specialist staff was
disbanded, and its monitoring role ended.
    With the termination of the management
contracts, RCM's and NCCM's Managing Directors
became government appointees and their Technical
Directors and other senior technical staff were
seconded from AAC and AMAX rather than working
directly for those companies. The earlier outline of
decision-making in the field of investment policies
indicates that these changes probably had little
significant impact. As mentioned above, the Managing
Directors can only make investment decisions on the
basis of the Technical Directors' recommendations.
The latter and their senior staff are usually long-
time employees of AAC or AMAX/RST, and many hope to
achieve further advancement within these
companies.[78] Without questioning the integrity of
the individuals involved, it is very likely that
this background influences their perceptions and the
criteria they apply in making decisions. Thus an
independent source of expertise was still required
if government directors were to ensure that criteria
other than those employed by the minority
shareholders could be applied in decision-making.
    A number of cases have arisen since 1974 which
illustrate this point. One resulted from the
government's desire to establish a fertiliser plant
at Kafue to reduce Zambia's dependence on imports.
Additional domestic supplies of sulphuric acid were

required if this project was to proceed. Collection
of sulphur emissions from the Mufulira smelter would
provide these supplies, and the government asked
RCM's technical staff to examine the feasibility of
installing a collection plant. Their study concluded
that technical and economic factors would render
such a project infeasible.[79] This may or may not
have been the case - the important point is that the
Zambian government lacked the capacity to obtain a
second opinion on the matter. It may have been that
the assumptions employed in the study were too
conservative, or that its terms of reference were
inappropriate from the government's viewpoint. For
example, the investment might not be warranted in
terms of its direct financial yield, but might be
desirable from a national perspective if the direct
and indirect effects on employment and the balance
of payments and the fact that it would considerably
reduce pollution from the smelter were taken into
account. The ability to ensure that broader
perspectives of this type are considered surely lies
at the heart of any attempt to establish national
control over mineral development. Other relevant
cases have involved the possibility of substituting
domestic coal for imported heavy fuel oil as a
source of power in certain processes, and that of
assisting small miners by giving them access to
smelting and other facilities. These are matters
which have been raised by the government[80] - it is
not known whether projects proposed by the operating
divisions have similarly been declared unviable.
What explains the government's failure to
retain the expertise developed in 1970-4? It may
have failed to realise the extent to which the
minority shareholders would retain their influence,
though MINDECO certainly understood the reality of
the situation.[81] However, by 1973 MINDECO's advice
was no longer being heeded; the organisation had
become involved in a political power struggle, and
this was in fact a key element in what occured.
MINDECO was headed by an extremely competent and
politically-astute Zambian, D.C. Mulaisho. Mulaisho
was deeply distrusted by Zambia's 'old guard'
nationalists, who saw him (correctly) as a potential
leader of an emerging political grouping of well-
educated young Zambians. This group favoured a more
pragmatic, economically-oriented approach to major
policy issues (especially in regard to Zambia's
relations with Rhodesia and South Africa), and it
was regarded as a threat to their dominance by the
older politicians. MINDECO, with its considerable

potential influence on the enormously important mining sector, represented a powerful political base for Mulaisho, and with its predominantly white expatriate staff was consequently suspect in the eyes of many senior politicians. Thus by 1973 MINDECO had powerful enemies who were glad of the opportunity to declare its activities superfluous, and who would have regarded claims that its continued existence was essential for effective government control of the copper mines as mere self-justification. This raises an important point - Zambia was still without a body of experts trusted by her senior politicians. MINDECO had been unable to stay above politics, while the Mines Department was still characterised by the weaknesses discussed in the previous chapter, except that turnover among senior Zambian staff was even more rapid after 1970,[82] and that many of its technical and professional positions were now simply unoccupied for want of suitable candidates.[83]

It is evident that majority ownership alone would not bring about effective government control, but was it essential if such control were to be established? For instance, the government might have gained considerable influence by obtaining minority participation (with the right to appoint a minority of directors) and developing sources of information and technical expertise on the lines described above. The latter would enable it to discover when corporate behaviour was in conflict with national policy aims; where this was the case, government directors could press for changes, and if their efforts failed sanctions could be imposed on the companies using normal legislative powers. Only in the area of profit reinvestment might such an approach leave the government without influence. Another alternative did of course exist in this area also - the government could have attempted to devise a taxation system which would have <u>encouraged</u> reinvestment.[84]

Another aspect of the Take-over involves its effects on the quality of middle-level and senior management in the industry. Many expatriates are not prepared to commit their careers to government-owned mining companies operating in only one country. The result is that turnover among middle-level professional, technical and managerial personnel is very high, while the calibre of recruits has fallen.[85] Disturbing in itself, the longer-term implications of this situation are even more serious. As senior technical and managerial staff leave the industry through retirement or for other reasons, suitable replacements will become more and more difficult to find. RCM and NCCM can, of course, attempt to recruit experienced personnel from abroad, but, particularly in the technical field, experience obtained elsewhere is a poor substitute for knowledge gained by working for many years in a particular industry and indeed frequently in a particular mine. Thus Zambia's copper industry may well face a managerial crisis until such time as Zambian graduates with sufficient experience become available.

Overview

Two points arise from this discussion. First, a rational analysis of its situation would have led the Zambian government to seriously consider alternative strategies to that of assuming majority ownership. Second, if it was decided that majority ownership was the only acceptable policy, such an analysis would have led the government to acquire and retain an expert knowledge of the copper industry. It did not adopt such an approach, primarily because of political factors, not simply short-term considerations of the type which determined the timing of the Take-over, but more fundamental factors which reflected basic features of Zambia's internal and international political situation.

As mentioned above, Northern Rhodesia was created through amalgamation of disparate tribal and linguistic groupings. By Independence, Zambians had not developed a common culture or language which might have acted as a unifying force, and most lacked any strong sense of national identity. Their primary identification was still with their tribe and the region it inhabited. Reflecting this situation, political allegiance followed tribal or regional lines. There was little support for the concept that a government, whatever its own political power base, should continue to distribute economic and other benefits to all Zambians regardless of tribal or geographical origin. Consequently loss of political power was equated with loss of rights and privileges, with the result that the political system has been under constant and severe strain from the threat of violent conflict between tribal and regional groupings, or of secession by one or other grouping. The continued operation of the system has depended on the ability of President

Kaunda, who is not identified closely with any one group, to centralise considerable executive power in his own hands, and to maintain a balance of power between the various factions.

This situation has affected the formulation of mineral policy in a number of ways. First, President Kaunda views his own role as crucial, and is consequently prepared to go to considerable lengths to maintain his position; this attitude explains, for example, his use of the 1969 Take-over decision for political purposes. Second, centralisation of power in the President's hands has led to a quasi-dictatorial method of decision-making. It will be remembered, for instance, that the President did not inform even his Cabinet of the Take-over decision before it was announced. This approach militates against the (sometimes extended) discussion and analysis which can be crucial to rational decision-making. Third, the President is constantly on guard lest any individual establish a power base which might threaten the delicate balance of power or his own position, while senior politicians are determined that their rivals will not steal a march. Fears of this type lie behind the constant shuffling of senior political and administrative personnel, and they underlay MINDECO's failure to survive as an efficient monitoring agency. Fourth, that sense of insecurity has made Zambian politicians extremely sensitive to powerful foreign interests in their midst, and no foreign interest was more powerful than the mining companies.

The companies were doubly feared and disliked because they were owned and managed by whites. It is important to emphasise again that many of the individuals who have governed Zambia since Independence sufferd insult and injury under a discriminatory political regime whose existence was inextricably linked with the activities of the copper companies. This had a major impact on governmental attitudes – it made it difficult to assess rationally the need for the companies, their skilled personnel, and the need for white advisers to assist in policy formulation. It was easy and indeed satisfying to take over the companies, dispense as quickly as possible with expatriate staff, and formulate policy without consulting outsiders. It would have been very much more difficult, and perhaps politically dangerous, to offer corporate tax concessions, encourage expatriates to stay, and take into account the advice of white advisers.

Political circumstances in Southern Africa reinforced these tendencies. The racism from which Zambians had suffered was still being practiced against fellow Africans in Rhodesia and South Africa. Objectionable in itself, this situation was made even more so by Zambia's vulnerability to economic and even military attack from these countries. And the mining companies were closely linked with those same countries. AAC was South African owned, both parent companies operated in Southern Africa, and both operating groups drew a high proportion of their goods and services and their employees from that region. Little wonder that Zambian leaders feared and distrusted the mining companies, or that they had a sharp desire to assume ownership of the copper industry, regardless of possible economic costs and indeed regardless of a rational examination of the best available means to acquire effective control.

Notes

1. RST International merged with AMAX in 1970.
2. RCM, 'Statement to Shareholders, 31 October 1975,' pp. 3-6.
3. RCM, Annual Report, 1972, p.10, NCCM, Annual Report, 1973, p.14.
4. In 1972/3, for example, NCCM's tax rate of 24 per cent was charged on profits of K100 million.
5. Republic of Zambia, 'Address by His Excellency, Dr K.D. Kaunda, at the Press Conference on the Redemption of the ZIMCO Bonds, 31 August 1973.'
6. This figure equals 0.75 per cent of the gross value of metal sales by RCM and NCCM.
7. Interview with Department of Finance Official, Lusaka, 1 August 1978.
8. Information could not be obtained regarding these items, but they are certainly significant.
9. Zambia's Financial Reports show that huge discrepancies frequently exist between budget estimates of mineral revenues, compiled early in the financial year, and revenue actually received in the same year. See, for example, Financial Report, 1971, Schedule B, and 1974, Schedule B.
10. Derived from Mining Journal, 24 February 1978.
11. The 1978 Budget provided for a decline in money terms of 4 per cent in current expenditure and of 14 per cent in capital expenditure.
12. At least two of AAC's wholly-owned subsidiaries manufacture mining equipment and supplies; AMAX does not appear to hold interests in any companies which

do so. Anglo American Corporation of South Africa Limited, Annual Report, 1976, pp.82-3; Mining Journal Limited, Mining Annual Review, 1979, pp.17-19 ('AMAX').
13. For example, there was a net loss of 32 expatriate metallurgical engineers over the period 1973-7, while only 10 Zambians were available to replace them.
14. As Table 4.1 indicates, copper production fell substantially between 1973 and 1978.
15. In 1978, turnover among expatriates was 35 per cent. Zambia Mining Year Book, 1978, p. 32.
16. These were among reasons cited by expatriate staff interviewed at RCM's Mufulira Division and NCCM's Chingola and Centralised Services Divisions, July-August 1978.
17. Derived from information supplied by MIMSU.
18. Shaft-sinking, development of main drives and cross cuts, and stope preparation were all at a lower level in 1977 than in 1970. Zambia Mining Year Book, 1970, p.32; 1978, p.30.
19. Interview with Department of Finance Official, Lusaka, 21 July 1978.
20. The industry has certainly encountered difficulties in maintaining production, despite substantial capital investment, but this has been due largely to shortages of goods and services and of skilled labour, and in most cases unskilled or semi-skilled labour could not be substituted for the missing factors.
21. For example, small scale productive units of the type operated by Africans in the pre-colonial period could not have competed with the giant open-pit mines of North America and Chile.
22. Baldwin, Economic Development and Export Growth, Chapter 4.
23. Prices for fabricated copper products fluctuate much less severely than those for copper metal.
24. See, for example, R. Prain, 'Prospects for Fabricating Copper in East and Central Africa', Selected Papers, 4 (The RST Group, London, 1968), pp.39-49.
25. In 1956/7, they accounted for over 80 per cent of domestic expenditures. See Baldwin, Economic Development and Export Growth, Table 2.4, p.38.
26. For example, the plant at Kafue which produces ammonium nitrate for use in making explosives has an annual output of about 26,000 tons, whereas annual production of about 50,000 tons is required to achieve minimum economies of scale. B. De Gaay Fortman, 'Zambia's Markets: Problems and

Opportunities', in Elliott, Constraints on the Economic Development of Zambia, p.217.
27. Ibid., pp.217-8.
28. Conflicts have occurred regarding the origin of those inputs which continue to be imported.
29. Sklar, Corporate Power, p.145.
30. Derived from NCCM, Annual Report, 1974, p.14 and information supplied by MIMSU.
31. Interview, Lusaka, 31 July 1978.
32. Derived from N. Kessel, 'Mining and the Factors Constraining Economic Development', in Elliott, Constraints on the Economic Development of Zambia, p.261.
33. Derived from NCCM, Annual Report, 1975, p.17.
34. It has been estimated that expatriate employees remit 30 per cent of their incomes. Derived from Goodman in Elliott, Constraints on the Economic Development of Zambia, p.235, n.4.
35. Total interest payments were K4 million, while about half of NCCM's loans were raised abroad. Interest rates on foreign and domestic loans are apparently similar.
36. Taxation was K164 million and the government's share of dividends K34 million. NCCM, Annual Report, 1974, pp.14, 17.
37. Derived from NCCM, Annual Report, 1974, and information supplied by MIMSU.
38. These figures were calculated on the same basis as those for 1973/4, from data contained in NCCM's 1976 Annual Report and information supplied by MIMSU.
39. Derived from NCCM, Annual Report, 1977, p.24, and RCM, Annual Report, 1977, pp.19-20.
40. NCCM, 'Circular to Members and Notice of Extraordinary General Meeting, 19 September 1978,' p.3.
41. This figure applies to the period 1971-6. Derived from Table 5.6 and Republic of Zambia, Economic Report, 1976, Table 6, p.27.
42. Martin, Minding Their Own Business, p.173.
43. This applies particularly to long-time residents of Zambia, who do not wish to live elsewhere, and to those nearing the end of thir working lives.
44. Interview with Mill Superintendent, Mufulira Mine, Mufulira, 26 July 1978.
45. For example it was 1973 before the University of Zambia could offer a complete range of courses corresponding to the mining industry's requirements.
46. MIMSU, 'From African Advancement to Zambianisation in the Mining Industry: Policies, Programmes, and Progress,' Mimeo, 1 June 1978, p.6.

47. Ibid., p.11.
48. Derived from Ibid., p. 14; MIMSU, Monthly Statistical Report, June 1978, Table A9.1.
49. Interview with Director, Anglo American Central Africa Limited, Lusaka, 1 August 1978.
50. This was the phrase used by the Department's Assistant Permanent Secretary to describe its approach to Zambianisation. Interview, Lusaka, 18 July 1978.
51. The records of the Manpower Planning, Training, and Zambianisation Unit show that, on average, Zambian graduates are promoted more rapidly than Europeans recruited at the same level. Interview with the Head of the Manpower Planning, Training, and Zambianisation Unit, Kitwe, 28 July 1978.
52. Ibid.
53. A major initiative was taken in this regard in 1973/4 when the number of expatriates engaged full time in training was doubled.
54. Alternative Development Strategies for Zambia (Land Tenure Centre, Wisconsion, June 1973), pp.5-6.
55. See, for example, Republic of Zambia, Report of the Auditor-General, 1974 (Government Printer, Lusaka, 1974), pp.8, 9, 14, 27.
56. In 1971 four government ministers were dismissed after a commission of inquiry found them guilty of financial improbity. In 1973, a number of senior politicians apparently used their access to confidential information to make hefty profits on trading in ZIMCO bonds.
57. The most serious of these involve unemployment, urban crime, and the poverty and disease of the shanty-towns which have sprung up in Lusaka and other centres.
58. I have documented this fact in detail elsewhere. See O'Faircheallaigh, 'Foreign Investment in Mineral Development,' pp.252-3.
59. Evidence of the industry's role as a 'wage leader' is presented by J.B. Knight, 'Wages and Zambia's Economic Develoment', in Elliott, Constraints on the Economic Development of Zambia, pp.99-102.
60. Richard Jolly has calculated that mineral revenue over the period 1965-9 would have been some K50 million greater if African wage increases had averaged 5 per cent per annum and expatriate rates had remained constant. R. Jolly. 'The Seers Report in Retrospect', African Social Research, 11 (June 1971), p.14.
61. Capital expenditure had to be reduced after periods of depressed prices in 1971/2 and 1975/6.

62. In the mid 1960s, wage payments accounted for from 20 to 40 per cent of total production costs in these sectors. Knight in Elliott, Constraints on the Economic Development of Zambia, Table 4.4, p.104.
63. Devaluation of the Kwacha might have counteracted these effects to some extent by making exports cheaper and imports dearer. However, mineworkers would probably be able to win compensatory wage increases, thus beginning the whole cycle once again.
64. Republic of Zambia, Second National Development Plan, p.9.
65. Sklar, Corporate Power, p.120.
66. A. Young, Industrial Diversification in Zambia (Praeger, New York, 1973), p.230.
67. Figures derived from Martin, Minding Their Own Business, Appendix, Fig. 7.
68. GDP increased at an average rate of 13 per cent between 1964 and 1970.
69. Derived from Martin, Minding Their Own Business, Appendix, Fig. 7.
70. Republic of Zambia, First National Development Plan, 1966-70, Ch. vi; Second National Development Plan, p.85.
71. Interview with L. Gustafson, Professor of Economic Geology, Research School of Earth Sciences, Australian National University, Canberra, 4 April 1978. It should be noted that the taxation system applied to mining companies encouraged high-grading by making exploitation of lower-grade ore unprofitable.
72. For example, much of the equipment on the Copperbelt was South African, and it might therefore be difficult to obtain spare parts elsewhere.
73. Interview with NCCM's Managing Director, Lusaka, 31 July 1978.
74. Interview with officials at NCCM's Centralised Services Division, Kitwe, 25 July 1978.
75. Interview with NCCM's Managing Director, Lusaka, 31 July 1978.
76. Interview with MINDECO's Consulting Engineer, Lusaka, 17 July 1978; interview with Director, Anglo American Central Africa Limited, Lusaka, 1 August 1978.
77. Interview with NCCM's Managing Director, Lusaka, 31 July 1978.
78. This applies especially to NCCM's senior staff, but in 1978 RCM's Technical Director, for example, was also on secondment.
79. Interview with MINDECO official, Lusaka, 17 July 1978.

80. Ibid.
81. Ibid.
82. Between 1970 and 1978, the Ministry of Mines had eight different Permanent Secretaries and the same number of Ministers. In 1978 alone, three different individuals held the post of Permanent [sic] Secretary.
83. In 1974, for example, only half of the established positions at this level were filled.
84. It is of course debatable whether such a strategy would have succeeded in the wake of the decision to take over the mines.
85. In fact many positions at this level are occupied by recent graduates seeking experience before applying for positions elsewhere.

Chapter Six

AUSTRALIAN POLICIES ON FOREIGN MINING INVESTMENT,
1945-1975

Australia was discovered by the Dutch in 1606,
declared a British colony in 1788, and became a
federation within the British Commonwealth in 1901.
Its land area is 7.69 million square kilometers,
much of it arid, barren, and remote from the eastern
and southern coastal cities in which most of
Australia's 14.3 million inhabitants live. Seventy
one per cent of its workforce are employed in
services (including government and public
authorities), and 21 per cent in a manufacturing
sector which enjoys a high degree of tariff
protection.[1] Agriculture and mining employ
relatively few Australians (7 and 1 per cent of the
workforce respectively),[2] but both are efficient,
export-oriented sectors which together provide
Australia with 75 per cent of its export income.
    Australia's federal system was established with
the intention of having a weak federal government
exercising enumerated powers in areas such as
defence, external affairs and immigration, and
strong state governments exercising all residual
powers and 'supreme in most fields of social and
economic concern'.[3] In the event, the federal
government's effective authority increased steadily
after 1901, with the two world wars and the
depression of the 1930s representing watersheds in
this process. However the states have vigorously
defended their traditional powers; of particular
significance in the present context is their
continued control over mineral rights, a factor
which has exercised an enormous impact on the
process of mineral development.
    The Liberal/Country Party governments which
have dominated federal politics since the late 1940s
have generally opposed government intervention in
commercial activity, except where private initiative
has been lacking. The Labor governments of 1972-5,

the second of which was removed from office after a major constitutional crisis, adopted a more inter-ventionist approach in this regard.

Australia has a long history as an important mining country, but at the end of the Second World War mining activity was at a low ebb. Over the next thirty years, this situation changed dramatically, and by 1975 Australia was one of the world's leading mineral producers. The first significant discoveries in the post-war period were of uranium, at Rum Jungle in the Northern Territory (1949) and Mary Kathleen in Queensland (1954); at the same time, the value of mineral sand deposits along the New South Wales and Western Australian coasts was recognised. These discoveries acted to stimulate exploration, and in the mid and late 1950s massive bauxite deposits were located in Queensland, the Northern Territory and Western Australia. The removal of export bans on iron ore in 1960 led to the discovery and development of huge reserves of the mineral in the Pilbara region of Western Australia. The mid and late 1960s saw discoveries of nickel in Western Australia and Queensland, uranium in the Northern Territory, and massive coal deposits in Queensland and New South Wales. In addition, production of 'older' metals such as copper, lead, zinc and tin expanded significantly. Over the period 1965-77, the value of minerals produced in Australia increased by more than four times (at constant prices), and the value of mineral exports by more than seven times.[4]

A number of factors accounted for this rapid growth in mineral exploration and development. Technological developments in the industrialised countries created markets for 'new' metals such as aluminium, nickel and uranium, while economic expan-sion in the United States, Europe and Japan greatly increased demand for minerals such as iron ore, coal and copper. The rapid growth of the Japanese economy was particularly significant, as Australia's geog-raphical location gave it a competitive advantage in supplying Japanese markets. Developments in mining techniques and in sea transport of bulk minerals facilitated the economic working and shipment of bulk commodities with low unit values (iron ore, coal, bauxite). Government policies also played an important part. The establishment of the Bureau of Mineral Resources, Geology and Geophysics by the federal government in 1946 provided exploration companies with a valuable and inexpensive supply of basic geological information. Tax concessions offered by the Commonwealth government represented a major financial incentive, and also demonstrated a generally favourable attitude towards mineral development, whether financed by Australians or foreigners, an important consideration at a time when many newly-independent mineral producing countries were enforcing harsher terms on foreign investors.

Many of the most significant and lucrative mineral developments of the 1950s and 1960s were financed and owned by foreign investors. This Chapter examines the policies adopted by Australian governments towards those investors, particularly over the period 1972-5. The Labor governments which held office during these years were notable for the number, range and significance of the initiatives they took towards the mining industry. However, their policies were rarely expressed or justified in a cohesive or comprehensive form; consequently, major policy initiatives will be examined individually, after which an attempt will be made to draw various policy elements together and to outline their underlying assumptions. In the next chapter, the validity of those assumptions and the appropriateness of government policies will be analysed in the light of an examination of the economic and other effects of foreign-financed mining projects.

## Australian Policies towards Foreign Investment, 1945-72

By the late 1960s, the extent of foreign ownership and control in the mining industry was causing concern to many Australians. According to Australian Bureau of Statistics (ABS) figures, foreign ownership in the metal mining industry (expressed as a percentage of the value of production accounted for) increased from 39.8 per cent in 1963 to 53.6 per cent in 1968, while foreign control increased from 51.0 to 68.7 per cent over the same period.[5] Concern with foreign investment in the mining industry was part of a broader re-appraisal of the role of such investment in the economy as a whole. Since 1945, Australian governments and their electorates had generally accepted that foreign investment was in Australia's benefit, but by the mid-1960s this assumption was increasingly being questioned.

The leaders of the ruling Liberal/Country Party coalition publicly stated their sympathy with their constituents' fears, but generally failed to formulate policies which might have allayed them. As early as 1964, the Prime Minister, Sir Robert

Menzies, expressed his concern with the increasing dominance of foreign investment, but proposed no action to stem or reverse the trend; rather he placed his hopes on an appeal to foreign investors 'to give Australians a share in their companies'.[6] In 1969 his successor, Mr Gorton, stated his government's desire to see foreign investors operating on a joint basis with Australian concerns and expressed confidence that overseas companies would respond to this desire.[7] Despite evidence that these appeals were falling on deaf ears,[8] similar statements were still being made by the Liberal Prime Minister, Mr McMahon, in 1972.[9] In the intervening years, very little had been done to ensure that the government's 'hopes' would be realised, while the extent of foreign ownership and control in the mining and manufacturing sectors steadily increased.

Ad hoc measures were taken in response to particular situations. In May 1965 restrictions were placed on local borrowings by foreign firms in an attempt to ensure that Australian companies would not be deprived of capital, but these were not rigorously applied. The government acted on two occasions to prevent foreign take-overs of individual firms in 'sensitive' areas (life assurance and uranium mining), and in 1972 an agency was established to screen future foreign take-over proposals, after a rush of such transactions had generated widespread public concern.

None of these measures represented a fundamental departure from the previous 'open door' policy towards direct foreign investment, and there was in fact no serious re-appraisal of policy in this area, reflecting the coalition's belief that the overall impact of foreign investment was overwhelmingly in Australia's favour.[10] This general attitude was reflected in the government's approach towards foreign investment in mining. In a major statement on mineral policy in 1972, the Minister for National Development, Mr Swartz, stated that a prime objective was 'to ensure that Australians participate effectively both in the equity capital of natural resource projects and in their management'. However the Minister did not offer any suggestions as to how this objective might be achieved, and indicated that there was no intention of taking any action regarding the high current level of foreign ownership and control in the mining industry.[11]

In the immediate post-war period, the Labor Party generally shared these favourable attitudes

towards foreign investment. However by the early
1960s Labor Members of Parliament were expressing
disquiet with the increasing role of foreign concerns,[12] an attitude given considerable impetus by
the large-scale foreign-financed mineral developments of the 1960s. By 1972 the earlier bipartisan
approach had disappeared. The Labor Party still
supported foreign involvement in mineral development
officially,[13] but it also advocated active intervention by the Commonwealth to contain the growth of
foreign ownership and control of mineral resources
and to encourage Australian participation in mineral
development.

In replying to the 1972 Ministerial statement
on resources policy, Labor spokesmen outlined their
party's position in greater detail. They recommended
that in future priority be given to Australians and
Australian companies in resource development. Where
mineral resources were already in the possession of
foreign companies, Australians should be offered a
chance to participate in their development either
directly or through the share market. If domestic
private investment were not forthcoming, the
Australian government should consider entering into
partnership with the foreign concern. In general,
they were critical of what they saw as the Liberal
Party's 'colonial mentality' and its 'soft approach
to multi-national corporations', and recommended
that government adopt a harder line to ensure
'maximum Australian equity, participation and
control'. More specifically, they argued the need
for government action to ensure that pricing
policies adopted by foreign mining concerns were in
Australia's interests, that such companies would
supply the government with detailed information
regarding their operations, and that Australian
minerals would, as far as possible, be processed in
Australia.[14]

On 20 September 1972 the Labor Party released a
statement of its policy towards foreign investment
in general. This emphasised the need to contain and
even reverse the trend of growing foreign ownership
and control in resource industries, stating that a
Labor government would 'act as necessary to retain
and regain maximum Australian ownership of and control of industries and resources'. It also outlined
concrete proposals for encouraging Australian participation in resource development. The role of the
Australian Industrial Development Corporation would
be expanded, with the objective of ensuring
'majority Australian ownership and control over

existing and future enterprises'. A National Development Corporation would be established to mobilise Australian capital by issuing Commonwealth-guaranteed National Development Bonds; where necessary, domestic funds would be supplemented by selling these bonds overseas. Another aim of this Corporation would be to ensure that existing Australian equity in resource projects remained in Australian hands. The government would also encourage life assurance companies and superannuation funds to increase their investments in resource projects by granting taxation and other incentives.[15] In general, the emphasis would be 'not so much [on] permitting foreign capital to do less [than on] permitting, persuading and enabling Australian capital to do more'. A Labor government would review the taxation provisions applying to mining companies, take action to ensure that export prices were maximised, and pursue maximum treatment and fabrication in Australia of Australian resources.[16]

The Labor Party was elected to office on 2 December 1972, and over the next three years attempted to implement the policies outlined above. For the purposes of this analysis, its initiatives can be divided into five categories, though many of the measures taken were in practice closely linked. The first category involved plans to mobilise Australian private capital, the second concerned direct government participation in mineral exploration and development, the third involved formulation of ownership guidelines for future mining projects, the fourth the application of controls to mineral exports, and the fifth the amendment of income tax legislation governing the mining sector.

## Mobilisation of Domestic Capital and the AIDC

The Australian Industrial Development Corporation (AIDC) was chosen as the agency through which it was hoped to mobilise private domestic investment capital for resource exploitation. Under its establishing legislation,[17] the AIDC's role was primarily to provide short-term assistance to Australian private enterprise in an attempt to increase domestic participation in industrial and resource development. It did this mainly by bringing together parties who in combination might be able to undertake major projects, and by providing a part of capital requirements (in loan or equity form) where this would ensure project viability. It was expected to sell any equity it acquired to private investors at the earliest opportunity, it could only purchase equity when providing capital for development, it could not itself initiate investment proposals, and could not be the sole Australian participant in joint ventures with foreign concerns. Its borrowing powers were limited and it was normally expected to borrow abroad. Clearly the AIDC's powers and the scope of its activities would have to be expanded if it was to perform the role assigned to it in Labor Party policy.

Proposals to do so were incorporated in legislation introduced by the Minister for Overseas Trade and Secondary Industry, Dr Cairns, in August 1973. The Australian Industry Development Corporation Bill 1973 proposed that the AIDC would be allowed to initiate investment proposals itself, retain shares it purchased, borrow more substantial amounts on foreign and Australian markets, and purchase equity in existing concerns. Capital to fund the AIDC's activities would be mobilised through the National Investment Fund (NIF)[18] to be established by concurrent legislation. The NIF would set up savings plans similar to superannuation schemes to which individual Australians could contribute; contributions up to a certain limit would be tax deductible. It would also establish a number of investment divisions, and for each division there would be a series of investment bonds and a corresponding portfolio of assets. Income to the investor would be paid out of the income earned by the assets in his division of the fund, and to encourage investment the government would contribute $10 to the fund for each $100 invested by individual Australians. Investment bonds would also be used to channel foreign investment into mining projects. The foreign investor would be entitled to income earned by the project and to the benefit of any growth in its capital value, but ownership and control would remain in Australian hands.

The AIDC's principal aims would be to create opportunities 'for the ordinary citizen to share in the kind of capital appreciation which so far has been confined to a very small percentage of the Australian people', and in the process to strengthen the individual Australian enterprise so that large Australian companies could grow up alongside the existing foreign ones.[19]

The AIDC would have two other roles. Where the Corporation deemed it 'economically prudent and possible to do so' it would act to regain control of foreign-owned industries. Indeed in Dr Cairn's personal view one of the AIDC's main aims would be to 'buy back Australia',[20] and he stated his approval for a proposal by Conzinc Riotinto of Australia (CRA) to sell 5 per cent of its equity to the Corporation.[21] The second role would be to act as an intermediary for government involvement in mineral-related developments. Where the AIDC decided it should not invest contributors' funds in a project because of high risk levels or the scale of expenditures involved, the government might decide to provide the necessary capital itself, and would do so through another division which it proposed to add to the AIDC, the National Interest Fund. Labor spokesmen made it clear that the Fund would be used as a vehicle for direct government participation in resource development, particularly in mineral processing.[22]

The Liberal and National Country parties opposed the Bills on a number of grounds.[23] They claimed that the AIDC would do little to mobilise additional domestic investment, pointing out that Australia already had a very high savings ratio by world standards. Rather, because of the incentives they would offer to investors, the AIDC and the NIF would successfully compete for funds with existing savings institutions and with government and semi-government agencies. Capital invested by the Corporation would thus be withdrawn from other uses, for example provision of housing and other infra-structure. If the foreign investment displaced by the AIDC left the country, the overall level of investment would decline, with consequent loss of income, employment and government revenue. The opposition argued that in fact foreign concerns would not invest in enterprises which they neither owned nor controlled, and would not transfer their funds into activities traditionally carried on by building societies and other saving institutions.[24]

Opposition spokesmen were particularly critical of schemes to 'buy back' equity in foreign mining companies, arguing that they might not in fact ensure Australian control of the enterprises involved while creating costs of the type outlined above. Support for this argument was drawn from the 1972 Treasury Paper on Overseas Investment, which had outlined the manner in which foreign mining companies could, by well-timed minority issues of shares in their subsidiaries, increase the size and profitability of their Australian operations without foregoing ultimate control. The Treasury had concluded that 'such issues may in fact yield a rather small return to Australia in terms of alternative uses of domestic savings'.[25]

In addition, the opposition was not convinced that the AIDC could guarantee returns on funds higher than or even equal to those being earned by private investment corporations, and believed that it might well earn lower rates as its investment decisions would be influenced more by the desire to maximise Australian ownership than by purely commercial criteria. Consequently opposition spokesmen could see no reason for transfering private investment capital to the AIDC. They also claimed that Australian control of mineral development could be ensured without resort to what it saw as the unnecessary expense of acquiring ownership of foreign mining concerns.[26]

The government's view was that foreign concerns would in fact be willing to invest in mineral development through the AIDC, partly because such investment would yield a substantial profit, but more importantly because the investor would be given guaranteed access to the minerals being produced: this was thought to be a more significant consideration than being granted equity, particularly to Japanese companies operating in the energy minerals sector (coal, uranium).[27] Government spokesmen pointed out that Australia generated some 90 per cent of her capital requirements internally, presumably implying that the economy would not be unduly affected if foreign funds were withdrawn. It was also emphasised that a major part of the AIDC's activities would involve mobilisation of Australian capital to develop new mineral discoveries, and that in these cases there would be no question of displacing foreign capital.[28]

The government claimed that Australia would suffer considerable loss if it failed to bring foreign-financed mining projects under national control. Its principal concern was that mineral exports would be underpriced, reducing tax liability (and so government revenue) in Australia.[29] As Dr Cairns stated,

> without strong Australian participation, we will have the prospect of Australian resources and industries being developed to maximise the global profits of multinational corporations, and very often at lower prices than if we in Australia were able to match a little closer their monopolistic powers.[30]

More generally, it was claimed that Australia's lack
of control over her mineral resources was one factor
in placing her in an 'economically subservient
position' in relation to her principal trading
partners, denying her the opportunity to use her
possession of valuable resources as a bargaining
counter in diplomatic negotiations, and effectively
reducing Australia's control over its own destiny.[31]
   The AIDC and NIF Bills were delayed in the
opposition-dominated Senate, reintroduced in the
House, delayed once again by the calling of a
general election in April 1974, and introduced for a
third time in July 1974. The Senate finally accepted
an amended version of the AIDC Bill in March 1975.[32]
It continued to reject the NIF legislation, which
was of course integral to the government's whole
concept of the AIDC's role, and which had not yet
been ratified when Labor was removed from office in
November 1975.

Direct Government Participation in Mineral
Exploration and Development

The platform on which the Labor Party fought the
1972 election included a call for 'the comprehensive
development under Government control of Australia's
mineral resources with emphasis on the need for
discovery of new deposits and direct Commonwealth
and State participation in oil and mineral search
and exploitation...'[33] In December 1973 the Minister
for Minerals and Energy, Mr Connor, introduced
legislation for the establishment of a Petroleum and
Minerals Authority (PMA), which was intended to be
the principal vehicle for federal government action
in this area.
   The function of the PMA would be 'to explore
for and develop petroleum and mineral resources and
to assist in implementing the Government's policy of
promoting Australian ownership and control of
natural resources and resource industries'. A
crucial aspect of its work would be to help implement the government's energy policies, and consequently much of its effort would be concentrated on
petroleum and natural gas, but its activities would
extend to non-fuel minerals. In particular, the PMA
would provide assistance to Australian ventures
which had discovered promising mineral deposits but
were unable to finance further exploration and
development. Mr Connor pointed out that existing
Australian mining concerns had a limited capacity to

take on new ventures, and that local exploration
companies found themselves turning to foreign
investors to provide finance; as a result, the
latter secured an interest in Australia's mineral
resources.[34] The PMA would remove the need for
foreign capital by providing equity, loans, or
guarantees, financing its activities from funds
saved by the Treasury by abolishing tax concessions
previously granted to individuals and companies
contributing to private exploration programmes (see
below).[35] Mr Connor stated the PMA would not take
over foreign equity in existing mining ventures, but
only assist local concerns in developing new
discoveries.[36]

In pursuit of its functions the PMA could
'undertake activities appropriate to a petroleum and
mining business.... employing its own personnel and
equipment in search, letting out contracts for
search, acting in partnership with
companies,...[and] taking up shares in companies'.[37]
The Authority would initiate exploration on its own
behalf by publishing a notice in the government
Gazette to the effect that a particular area had
been declared an exploration area for the purposes
of the establishing Act. It would not enter land
until the written permission of the occupier had
been received, but if this was not forthcoming a
Justice of the Peace could issue a warrant
permitting entry.[38]

The opposition rejected the proposed legislation vehemently. It claimed that the powers of
entry conferred on the Authority would allow it to
usurp promising exploration areas held by private
concerns, take over mineral deposits discovered by
other prospectors, and even nationalise existing
mining operations. Undesirable in itself, this prospect would also have serious side effects. It would
destroy the traditional power of the states to
allocate mineral rights. The threat of nationalisation would have a disastrous effect on mineral
exploration. Foreign companies would be unwilling to
invest risk capital knowing that any discoveries
might be confiscated, while the uncertainty created
by the PMA's existence would make it difficult for
local concerns to raise finance. The opposition also
pointed out that the Treasury would save only about
$50 million by abolishing the tax concessions, a sum
which could not possibly make up for the predicted
decline in private exploration. In addition, it
opposed in principle the use of taxpayers money in
high-risk activity.[39]

172

In reply, government spokesmen argued there was a need for direct government participation in the minerals and energy sector if decision-making was to be based on the 'community good', and not just on the motivation of private shareholders or multi-national corporations. More generally, they claimed it was necessary to increase Australian ownership of resources (by public **and** private concerns) if Australia was not to be 'ripped and raped' by foreign companies. In particular, concern was expressed that Australian mineral exports would be underpriced while such companies were in control.[40]

The Petroleum and Minerals Authority Bill was rejected on two occasions in 1973-4 by the opposition-dominated Senate. Along with a number of other bills it was the basis for a double dissolution of Parliament in April 1974, and when the Labor Party was returned to office it pushed the measure through in a joint sitting of both Houses. The Act was already under challenge in the High Court on technical grounds, but the authority was constituted in any case, in August 1974. The High Court found the Petroleum and Minerals Authority Act invalid in June 1975. A considerably modified version of the Act was introduced in October 1975, but the Labor Pary's electoral defeat in December prevented further action.

During the short period for which it was operational, the PMA had available neither the time nor the resources to organise an exploration programme on its own behalf, but it did become involved in a number of existing mining ventures. For example in October 1974 it acquired a 49 per cent shareholding in the Wambo Mining Corporation, a New South Wales coal producer which was then suffering from cash shortages. Wambo unsuccessfully sought help from a number of Australian companies, after which it considered admitting the Anglo American Corporation of South Africa Limited as a shareholder under an arrangement which would have raised its level of foreign ownership from 17 to 53 per cent. Wambo approached the PMA, which agreed to pay $3.7 million for 49 per cent of the company and provide credit facilities. But the PMA's largest single venture was motivated by a desire to increase Australian ownership in a foreign firm. This involved the acquisition of a shareholding in Dehli International Oil Corporation, a wholly US-owned company which controlled oil and natural gas deposits in South Australia.

Another vehicle for direct government partic-

ipation was the Australian Atomic Energy Commission
(AAEC). In mid-1974 the AAEC agreed to underwrite a
share issue by the CRA subsidiary, Mary Kathleen
Uranium (MKU), when it became apparent that the
level of foreign ownership in MKU would otherwise
increase; as a result, the AAEC obtained a 46 per
cent holding in the company. The government
undertook a much more substantial commitment by
involving the AAEC in the Northern Territory Ranger
uranium project, controlled by two predominantly
Australian-owned companies, EZ Industries and Peko-
Wallsend. In October 1973 the government concluded
an agreement with these companies whereby it would
provide 72.5 per cent of the project costs in return
for 50 per cent of the uranium concentrates
produced. According to later reports, EZ Industries
and Peko-Wallsend were far from happy with the
government's involvement, which they regarded as
'effective expropriation', and only agreed to the
arrangement in the belief that to do otherwise would
be to risk losing their entire investment.[41]
What prompted the government to become involved
in an Australian mining enterprise? Its concern to
exercise control over energy sources was clearly
important; in addition, it may have felt that the
Ranger partners' need for finance might force them
to involve foreign investors in the project. Perhaps
most importantly, two of the other large Northern
Territory uranium deposits, which would be developed
later, were foreign owned (Jabiluka and Koongara).
The government had decided to acquire a 50 per cent
holding in these ventures,[42] and may consequently
have felt that it had to set a precedent with the
first major development.

Ownership Guidelines for the Mining Industry

The third category of initiatives involved the
formulation of ownership guidelines for future
mining projects. At an early stage the government
made it clear that in future there would be 'at
least majority Australian control over both equity
and policy in resource development'.[43] In October
1973 the Prime Minister, Mr Whitlam, explained the
government's position in greater detail. Its general
aim would be 'to achieve the highest possible level
of Australian equity that can be achieved in
negotiations project by project, that are fair and
reasonable to both parties and are within the
capacity of our own savings to support'.[44]
Thus Mr Whitlam indicated his government's

belief that continued foreign investment might be required, and he went on to say:

> I want to make it clear that there is no prescription of foreign equity participation in mining. In some circumstances it may be acceptable for foreign investors to participate significantly in decision-making in a project. The size of the project, the amounts involved, and the type of mineral are all factors to be taken into account. In pursuing our objectives we shall be flexible and guided by the particular needs of particular cases.[45]

However this flexible approach would not apply to fuel minerals, particularly uranium. Growing world energy shortages and 'other factors' dictated that full Australian ownership and control would be sought in this area.[46] Nevertheless the government recognised that in this case also Australia's resources of capital and technology might be inadequate, and that it would consequently 'need to call upon overseas expertise, technology and capital...'. However overseas participation would be sought 'through access to technology, loans and especially long-term contracts', rather than through equity ties.[47]
   Exploration for uranium would also be carried out exclusively by Australian concerns, but in regard to exploration for other minerals Mr Whitlam stated:

> In order to maintain a desirable level of exploration activity, we would, if necessary, accept a lower level of Australian ownership in exploration. Given the limited Australian capital resources available and the higher risks usually involved in exploration, there is much to be said for concentrating equity at the production stage.[48]

This policy was based on a number of important assumptions. The first was that foreign investors would be willing to supply capital and technology without obtaining equity. The second was that Australia could generate sufficient capital to maintain or increase exploration for fuel minerals, and to provide a high proportion of development finance for fuel and non-fuel mineral discoveries. The third was that foreign mining concerns would be willing to finance the bulk of expenditure on exploration for

non-fuel minerals in the knowledge that, if successful, they would be expected to offer an unknown
proportion of their equity to Australian concerns.
The validity of the first two assumptions is
analysed in detail below; as regards exploration,
time alone would tell whether foreign companies
would be prepared to operate under these
conditions,[49] but the view of overseas companies
then operating in Australia was that they definitely
would not.[50] One point was clear - the guidelines
relating to exploration and development of non-fuel
minerals left considerable uncertainty as to what
level of foreign participation would be acceptable,
and this in itself was likely to have an adverse
effect on exploration.[51]

Though accepting the need for foreign
investment in exploration, the government did insist
that Australian concerns be invited to participate
in all new exploration ventures. Over the period to
March 1975, it refused some 30 foreign companies
permission to import funds for exploration until
this requirement had been satisfied.[52]

More definite equity guidelines were not laid
down until September 1975. The requirement to seek
Australian participation at the exploration stage
was removed, and Mr Whitlam announced that:

> the Government will expect proposals for all
> new mineral development projects (other than
> those relating to uranium) to have no more than
> 50 per cent foreign ownership... The Government
> will not insist on the Australian participants
> necessarily being the technical operators and
> supervisors of the project, but will expect
> Australian nationals to have a significant role
> to play in the management, technical operation
> and control....In considering proposals for
> foreign equity participation...account will be
> taken of the expected main source of demand for
> the output of the development.

However it was stressed that all foreign investment
proposals would be examined on a case by case basis
to ensure maximum benefit to Australia.[53]

The reference to the source of demand for
mineral output reflected the government's recognition that availability of secure markets was
frequently dependent on the potential customer being
permitted to take a major shareholding at the mining
stage. This represented a fundamental departure from
the previous position, which was based on the

assumption that customers would be willing to forgo ownership in return for security of supply, and which ignored the fact that many consumers equated security of supply with a measure of ownership and control. Mr Whitlam's announcement was greeted with approval by mining industry spokesmen, who felt that the stipulation of minimum Australian equity removed much of the uncertainty created by the previous ad hoc approach, allowing foreign companies to plan accordingly.[54]

## Export Controls

Another means of implementing policies was provided by the Commonwealth's control over mineral exports, derived from the federal parliament's jurisdiction over international trade. Mineral exports had been subject to federal control on a number of occasions prior to 1972, mainly for purposes of conservation (iron ore from 1938 to 1960) or on grounds of national security (uranium). In the decade to 1972, controls were generally relaxed, though the government did intervene to insist on price increases for iron ore shipments to Japan in 1966, and to set floor prices for zircon exports in 1971. All of these actions were taken on an ad hoc basis in response to specific problems, and generally involved only one mineral at a time.

When the Labor Party came to office, it announced that all minerals would be subject to control on a continuous basis. Beginning in March 1973, applications for export licences would have to be cleared by the Department of Minerals and Energy. In order to obtain clearance, exporters would have to supply the Department with copies of existing sales contracts and other relevant documents affecting quantities, price and conditions of export, and to consult with it prior to finalising new export contracts.[55]

The government had at least three distinct reasons for applying export controls. The first was to obtain information. Mr Connor complained that his predecessors had kept no records regarding export prices obtained, the degree to which these corresponded to 'world parity' prices, the periods over which contracts applied, or the currencies in which they were denominated.[56] Second, the government feared that mineral exports might be undervalued, that transfer pricing might occur between subsidiaries of vertically-integrated multinational mining companies or that closely co-ordinated buyers might hold a bargaining advantage over Australian suppliers who were fragmented and frequently competing among themselves for export markets. Of particular concern in this latter regard were sales of iron ore and coal to Japanese companies, who operated through a single buying agency. Third, export controls could provide the government with a power of veto over the development of new mines, allowing it to ensure that its guidelines on issues such as foreign equity participation or environmental protection were adhered to.[57]

In 1973-4, the government utilised its export controls to impose a more co-ordinated approach on mineral exporters, particularly on Western Australian iron ore producers. Price guidelines were agreed upon at meetings with these producers prior to each set of negotiations with the Japanese, a strategy which apparently had some success. In May 1973 the iron ore companies won substantial price increases from the Japanese to compensate them for devaluation of the Australian dollar (their contracts were denominated in US dollars). In August 1974 Mr Connor instructed the producers to re-negotiate increases in iron ore prices, with the result that prices increased by 28 rather than 24 per cent.[58] Subsequently, the depressed state of world iron ore markets restricted the government's ability to act in this way, and no further increases were obtained until April 1976. Prices for coal exports also increased substantially after the controls were introduced, but this was apparently due to the impact of world energy shortages rather than to the government's intervention.[59]

Mining Industry Taxation

Under provisions of the Income Tax Assessment Act implemented in 1968 and prior years, mining companies and certain of their shareholders were permitted to reduce and/or postpone their liability to taxation by means not available to other taxpayers. These special provisions fell into four categories. Sections 77 and 78 allowed individuals subscribing to shares in exploration or mining companies to deduct part or all of the amount subscribed in calculating their taxable income. Section 23 exempted from taxation 20 per cent of net income earned from mining prescribed minerals, including bauxite, copper, nickel, mineral sands and tin. The remaining concessions, involving deduction of capital expenditure in assessing taxable income, were covered by Division 10 of the Act. 'Category A' concessions permitted deductions in respect of items not allowable for other taxpayers, for example preliminary site preparation and construction of housing and community facilities. 'Category B' concessions allowed mining companies to apply generally-available allowances at an accelerated rate. The effect of these concessions was that, especially during periods of major capital expenditure such as occurred in the mid and late 1960s, mining companies paid rates of tax considerably lower than those applied to all corporations, as Table 6.1 illustrates.

The concessions to individual taxpayers granted by Sections 77 and 78 were apparently inefficient in promoting mineral exploration. According to one estimate, only 27 per cent of the amounts claimed was actually spent on exploration or development, while the concessions were employed by some subscribers in tax avoidance schemes.[60] In May 1973 the Labor government announced their termination, claiming that exploration could be promoted more efficiently by channeling the funds saved by the Treasury through a government mining corporation.[61] Some months later the Minister for Minerals and Energy, Mr Connor, commissioned a consultant, T.M. Fitzgerald, to examine the whole question of income tax for the mining industry. His report, The Contribution of the Mineral Industry to Australian Welfare, was published in April 1974.[62]

Fitzgerald concluded that the application of the Income Tax Assessment Act represented a substantial subsidisation of the mining industry from public funds. He rejected industry claims that the concessions relating to capital expenditure simply postponed liability to taxation, pointing out that Category A deductions were not available to other companies and that their application exempted permanently from taxation a proportion of income which would have been taxable in the hands of such companies. He also pointed out that, where capital expenditure was maintained at high levels over an extended period of time, application of category A and B concessions would postpone liability to taxation long after expenditures associated with the initial development had been allowed for. By way of illustration, he quoted the example of Mount Isa Mines, which was then still paying less than 10 per cent of its net income in tax, twenty years after the initial development of its copper deposits was completed.[63] In fact, Fitzgerald argued,

Table 6.1: Tax Rates, Mining Companies and All Companies, Australia, 1967/8 - 1973/4

| YEAR | 1967/8 | 1968/9 | 1969/70 | 1970/71 | 1971/2 | 1972/3 | 1973/4 | TOTAL |
|---|---|---|---|---|---|---|---|---|
| All Companies | | | | | | | | |
| Company Income ($m)* | 2,219 | 2,586 | 2,972 | 2,862 | 2,996 | 3,709 | 4,206 | 21,550 |
| Income Tax ($m) | 938 | 1,070 | 1,337 | 1,367 | 1,448 | 1,885 | 2,408 | 10,453 |
| Tax Rate (%)** | 42 | 42 | 45 | 48 | 43 | 51 | 57 | |
| Mining Companies | | | | | | | | |
| Company Income ($m)* | 106 | 132 | 358 | 345 | 471 | 484 | 555 | 2,451 |
| Income Tax ($m) | 29 | 49 | 78 | 37 | 46 | 99 | 187 | 524 |
| Tax Rate (%) | 27 | 37 | 22 | 10 | 10 | 20 | 34 | 21 |

* Excludes tax payable by private companies
** Tax ratios vary from the company tax rate because the income on which tax is assessed by the taxation authorities differs from company income calculated for National Accounts purposes.
Source: Bureau of Mineral Resources, Geology and Geophysics, Australian Mineral Industry Annual Review, various issues; Industries Assistance Commission, Report, Petroleum and Mining Industries, 28 May 1976 (Government Printer, Canberra, 1976), Table 1, p. 53.

a mutually reinforcing relationship existed between the tax concessions and growth of productive capacity in the mining industry: growth created tax concessions, and the tax concessions helped to finance greater growth while also providing incentive for growth.[64]

Fitzgerald opposed the concessional treatment afforded the mining sector on a number of grounds. First, he claimed it might encourage inefficient allocation of resources, by subsidising investment in export generating activity which was surplus to Australia's foreign exchange requirements. He argued that the priority attached by policy-makers to earning foreign exchange in the 1960s was no longer appropriate, since Australia's balance of trade was in surplus and her foreign reserves substantial. The principal consideration in allocating resources should be the impact of alternative uses on national income and its distribution; since foreign ownership was high in the mineral industry, concentration of resources in that sector would result in a growing drain on national income, and was consequently not to be encouraged. Fitzgerald also feared that the economies of capital cost which the taxation system allowed an expansive and heavily-borrowing mining company might mean that it would benefit from increased production even where this resulted in lower prices. If this occurred, Australia might be denied part of the value of its resources, to the benefit of (predominantly foreign) buyers of the minerals concerned.[65]

His second principal objection was based on the contention that a high proportion of the value of the mineral industry's output derived from the intrinsic worth of non-renewable resources belonging to the Australian people. It was consequently essential that Australia should reap substantial financial benefit from exploitation of its minerals. According to Fitzgerald's calculations, the granting of tax concessions combined with the federal government's provision of services to the mining industry (for example through the BMR) resulted in a net financial **loss** to the Commonwealth in its relations with the industry, amounting to $55 million over the years 1967/8-1972/3. In comparison, foreign direct shareholders of the principal mining companies earned $1,024 million in profits, net of current taxation, during this period.[66]

Mining industry and other critics attacked the **Fitzgerald Report** on a number of grounds, both methodological and conceptual. It was claimed that the sample companies on which it was based were atypical of the industry as a whole, being without exception large, profitable enterprises, and that consequently its assessment of the industry's ability to bear additional taxation was invalid.[67] The companies chosen by Fitzgerald for detailed analysis were certainly amongst the most profitable in the industry, and he failed to substantiate his claim that mining was Australia's 'most profitable sector'.[68] In fact, over the period 1967/8-1973/4, profitability of Australia's 27 largest mining companies (measured by the average rate of return to funds employed before tax but after deduction of royalties) was only slightly higher than in the manufacturing sector (16.4 as against 12.6 per cent).[69]

However, this average figure hid very large disparities between individual mining concerns: some made losses, while others earned annual returns of up to 50 per cent on funds employed.[70] While any assessment of mining industry taxation should certainly have taken the existence of loss-making or marginal operations into acount, it does not necessarily follow that average profitability should determine the rate of tax to be applied to all companies. Government might wish to charge substantial rates on highly profitable operations while leaving others free from tax or subject to lower taxes, especially since a proportion of above-average profits might derive from the relative richness or accessibility of the deposits involved, indicating the existence of economic rents.

The Fitzgerald Report's claim that Category A allowances (those not available to non-mining companies) represented a concession to the mining industry also came under attack. The industry argued that these allowances related to essential capital assets which were either not used by non-mining companies (for example shafts, tunnels), or which would have no residual value in the remote regions where mining usually took place but would retain or even increase their value in urban areas where most non-mining companies operated. Consequently, it was claimed, the only concession offered by Division 10 derived from the fact that mining companies could write off assets more quickly than could other concerns.[71]

Mining industry spokesmen recognised the value of this concession,[72] but argued that the industry operated under unique disabilities and that some

recompense was necessary if it was not to be disadvantaged relative to other sectors. Among these
perceived disabilities were the necessity to undertake expensive exploration programmes, liability to
royalty payments, the need to provide infrastructure
usually supplied by public authorities, and a high
level of dependence on export markets characterised
by severe price fluctuations. In addition, the
higher cost structure imposed by remoteness meant
that operations had to be on a massive scale to be
viable, facing the industry with special problems in
raising huge blocks of development capital. This in
itself, it was contended, necessitated application
of accelerated depreciation allowances, since loan
funds would be difficult to obtain without the
assurance of large early cash flows.[73]
   Some of these arguments were certainly valid;
the mining industry did display distinctive
features, and it was appropriate that these should
be recognised in income tax legislation. However,
while the Fitzgerald Report may have exaggerated the
extent of the concessions available to the industry,
this did not preclude the possibility that unnecessarily generous concessions were being offered by
government in its attempts to encourage mineral
development.
   Another criticism of the Fitzgerald Report was
that it adopted too narrow a perspective in
assessing the mining industry's contribution to
Australian welfare in that it ignored, for example,
the industry's role in providing export receipts,
employment, and infrastructure, and in furthering
decentralisation and industrialisation (the latter
via mineral processing and fabrication). It was also
claimed that the narrowness of his approach led
Fitzgerald to misrepresent the nature of the
industry's financial relationship with government,
to ignore certain items on the positive side, for
example additional employee taxation resulting from
higher wage payments to mineworkers, and savings on
community infrastructure which public authorities
would otherwise have had to provide.[74]
   The Fitzgerald Report did not in fact ignore
the mining industry's direct economic impact, but
rather took the view that its significance had been
exaggerated. It claimed, for example, that mineral
development in remote areas such as the Pilbarra had
failed to bring about significant decentralisation,
and that at least certain types of mineral
processing (for example toll refining of bauxite at
Gladstone) had very little positive economic impact

on Australia. In general, Fitzgerald espoused the
view that 'if Australia is to gain much short-term
benefit from foreign exploitation of her mineral
wealth, it will have to be mainly through the taxes
and royalties imposed on overseas-owned mining
companies'.[75]
    The significance of the mining industry's
direct economic impact is examined in the next
Chapter, but at this stage one general comment can
be made. Regardless of how significant that impact
was, it was still pertinent to ask whether tax
revenue was being sacrificed unnecessarily. In other
words, would the mineral developments of the 1950s
and 1960s have proceeded without the tax incentives,
or with less generous incentives?
    Clearly this question cannot be answered con-
clusively. Some relevant evidence was collected by
the Industries Assistance Commission Inquiry into
the mining and petroleum industries, held over the
period 1974-6. The Inquiry heard evidence from most
major mining companies operating in Australia. No
company could provide evidence of any project that
would not have been undertaken had the concessions
not applied, and the Commission concluded

> that the major part of the development which
> occurred over the period 1967/8-1972/3 would
> have taken place in the absence of the
> concessions... [though] the rate at which many
> of these projects were developed would have
> been slower.[76]

However, this conclusion raises two important
issues, the first of which involves the perspective
of policy-makers in the 1950s and 1960s. Policy-
makers were apparently not fully aware of the
various factors which favoured Australia in seeking
development of her mineral resources (see above pp.
163-4), and felt that tax concessions were essential
if development was to proceed.[77] There seemed little
to lose and much to gain in terms of direct economic
impacts, which were expected to be substantial, by
offering the concessions. Even had there been a
greater awareness of Australia's inherent advantages, Australian governments anxious to expand
export receipts quickly would probably have been
prepared to sacrifice revenue in order to ensure a
more rapid pace of development,[78] though they could
hardly have foreseen what the extent of the sacrifice would be. In the event, the tax liability of
mining companies was reduced by an estimated $1,100

million over the period 1967/8-1972/3 alone;[79] figures are not available for earlier years, but a number of major projects became operational prior to 1967/8, and the total loss was probably well in excess of this sum.

Second, it must be remembered that four quite distinct types of provisions accounted for the reduction in tax liability, and that their significance varied. Exemptions of mining income accounted for $117 million.[80] It is unlikely that these exemptions exercised a significant influence in the post war period; they did not even apply to two of the principal minerals developed at this stage (iron ore and coal), and generally benefited long-established concerns such as Mount Isa Mines. Allowance of share payments against tax liability cost some $240 million. It is improbable that large developments would have been delayed in the absence of this concession, since only a fraction of the amounts involved were actually spent on exploration or development and since most of the major companies obtained the bulk of their funds overseas. Category A allowances accounted for $300 million. Absence of these provisions would have placed a substantial additional burden on mining companies, and might well have led to a slower pace of development and possibly to abandonment of some marginal projects. The remaining $445 million represented deferred tax liabilities resulting from instant asset write-off, with government's loss represented by the (considerable) notional interest on this amount. It is particularly difficult to reach definite conclusions regarding this concession. Somewhat less generous provisions would have reduced the loss to government and might not have retarded development, but lack of any concession would probably have done so.

We may therefore conclude that the Australian government need not have sacrificed revenue to the extent it did to achieve its objectives, and that the cost of its generosity was somewhere in excess of $350 million over the period 1967/8-1972/3 alone.

The Labor government accepted the **Fitzgerald Report's** conclusions, withdrawing the exemptions of mining income from taxation (with the exception of gold mining) and terminating the accelerated depreciation provisions.

Labor Party policies were based on the assumption that foreign ownership in the mining industry was creating certain costs for Australia, and in particular that foreign companies were engaging in pricing and marketing practices inimical to Australia's interests, were failing to develop viable opportunities for mineral processing in Australia, and were responsible for substantial outflows of income from Australia. In addition, it was felt that Australia's perceived lack of control over the operations of foreign concerns denied it the opportunity to fully utilise the potential diplomatic power conferred on it by its possession of valuable mineral resources. Further, it was assumed that Australia could not re-establish control or avoid the costs of foreign investment unless direct Australian participation in the mining industry, public and private, was very much increased. The Labor Party believed that this could be achieved without significant decline in the scale or efficiency of mining operations or in overall levels of capital investment. It also believed that mineral development would continue under substant- ially less generous taxation provisions. Imposition of more stringent provisions was essential if profit outflows were to be curtailed in the short term, and since the direct economic impact of mining projects was unlikely to be substantial.

In general, Liberal/Country Party governments operated on diametrically opposed assumptions throughout the post-war period. There was little to fear from foreign mining investment and any dangers could be dealt with through the normal exercise of government powers. In any case Australia did not possess the resources necessary to develop her minerals, and taxation incentives were required if foreign investors were to be persuaded to do so. The direct economic benefits created by mineral develop- ment would be sufficient recompense to Australia for permitting exploitation of her minerals.

An assessment of these contradictory viewpoints requires an empirical analysis of the impact of foreign-financed mining projects, and this is attempted in the next Chapter.

Notes

1. Far Eastern Economic Review, Asia 1980 Yearbook (Hongkong, 1980), p. 134.
2. Ibid.
3. J.D.B. Miller and B. Jinks, Australian Government and Politics (London, 1971), p. 38.
4. Bureau of Mineral Resources, Geology and Geophysics (BMR), Australian Mineral Industry Annual Review, 1977 (Government Printer, Canberra, 1977),

Table 10, p. 12; Table 12, p. 18.
5. Commonwealth Treasury, _Overseas Investment in Australia_, Treasury Economic Paper No. 1 (Canberra, May 1972), Table 18, p. 25.
6. Quoted in B. Fitzpatrick and E.L. Wheelwright, _The Highest Bidder_ (Lansdowne Press, Melbourne, 1965), p. xiii.
7. _Commonwealth Parliamentary Debates (CPD)_, House of Representatives, Vol. 65, 16 September 1969, p. 1382. All references to _CPD_ in this and the following chapter are to House of Representatives Debates.
8. See, for example, D.T. Brash, _American Investment in Australian Industry_ (Australian National University Press, Canberra, 1966), pp. 67-71.
9. _CPD_, Vol. 80, 26 September 1972, p. 1920.
10. Ibid., p. 1917.
11. 'Australia's Natural Resources: Ministerial Statement and Review', Parliamentary Paper No. 212 (1972), pp. 2, 15-17.
12. See, for example, the statements by Labor MPs, _CPD_, Vol. 35, 11 April 1962, pp. 1588-98.
13. CPD, Vol. 81, 26 October 1972, p. 3343.
14. Ibid., pp. 3347-8.
15. 'Foreign Investment and Associated Matters: Resolution by Federal Parliamentary Labor Party, 20 September 1972' (Canberra, 1972).
16. _CPD_, Vol. 80, 26 September 1972, pp. 1922-3.
17. The Australian Industry Development Corporation Act 1970.
18. The National Investment Fund Bill 1973.
19. _CPD_, Vol. 85, 30 August 1973, pp. 655-7.
20. Ibid., 29 August 1973, p. 515.
21. _Age_, 14 April 1973. Dr Cairns apparently soon lost his enthusiasm for the CRA scheme.
22. _CPD_, Vol. 86, 16 October 1973, p. 2176.
23. Ibid., pp. 2172-94.
24. Ibid., pp. 2162-7.
25. Commonwealth Treasury, _Overseas Investment in Australia_, p. 97.
26. _CPD_, Vol. 86, 18 October 1973, pp. 2343-4, 16 October 1973, pp. 2172-4.
27. Ibid., 16 October 1973, p. 2170.
28. Ibid., pp. 2186, 2190.
29. Ibid., pp. 2170, 2190.
30. Ibid., Vol. 85, 30 August 1973, p. 653.
31. Ibid., Vol. 86, 16 October 1973, pp. 2186, 2191.
32. The most significant amendments prevented the AIDC itself from raising funds except by borrowings, enjoined it to dispose of shares 'not necessary for the performance of its functions', and placed

restrictions on its ability to gain control over business enterprises.
33. CPD, Vol. 87, 4 December 1973, p. 4244.
34. Ibid., p. 4248.
35. Ibid., Vol. 88, 8 April 1974, p. 1155.
36. Ibid., Vol. 87, 4 December 1973, pp. 4245-6.
37. Ibid., pp. 4247-8.
38. Ibid., 12 December 1973, p. 4602.
39. Ibid., pp. 4601-30.
40. Ibid., Vol. 87, 13 December 1973, pp. 4607-20; Vol. 88, 8 April 1974, pp. 1154-6.
41. Far Eastern Economic Review, 18 January 1980.
42. 'Northern Territory Uranium Program Outlined: Media Statement, R. Connor, 31 October 1974'.
43. CPD, Vol. 83, 12 April 1973, p. 1411.
44. 'Foreign Investment in Australia: Speech by the Prime Minister, Mr Whitlam, at the Australian-Japan Ministerial Committee Meeting, Tokyo, 29 October 1973'.
45. Ibid.
46. Ibid.
47. Ibid.
48. Ibid.
49. In real terms, expenditure on mineral exploration fell substantially between 1972 and 1975, but it is not apparent whether this was due to the equity guidelines alone, to uncertainty created by the Labor government's other initiatives, or to other factors.
50. Age, 11 August 1973.
51. Ibid.
52. Age, 1 March 1975.
53. 'Policy on Foreign Investment: Statement by the Prime Minister, Mr Gough Whitlam, 24 September 1975'.
54. Age, 25 September 1975.
55. 'Mineral Exports Controlled: Press Statement, R. Connor, Canberra, 23 February 1973'.
56. CPD, Vol. 83, 3 April 1973, p. 969.
57. Ibid., p. 49.
58. Australian, 31 May 1973.
59. G. Stevenson, Mineral Resources and Australian Federalism, Centre for Research on Federal Financial Relations Research Monograph No. 17 (Australian National University, Canberra, 1976), p. 52.
60. R.B. McKern, 'Foreign Direct Investment in the Financing of the Australian Minerals Industry', paper presented to the 101st annual meeting of the A.I.M.E., San Francisco, 1972; Industries Assistance Commission (IAC), Report, Petroleum and Mining Industries, p. 47.

61. CPD, Vol. 83, 8 May 1973, p. 1750.
62. Government Printer, Canberra, 1974 (known as the Fitzgerald Report).
63. Fitzgerald Report, pp. 14, 32-7, 90.
64. Ibid., pp. 24-5.
65. Ibid., pp. 3, 64, 69, 71, 74.
66. Ibid., pp. 3, 74, Table 1, p. 6.
67. S. Bambrick, The Changing Relationship: The Australian Government and the Mining Industry (Committee for Economic Development of Australia, Melbourne, May 1975), p. 29.
68. Fitzgerald Report, p. 8.
69. IAC, Report, Petroleum and Mining Industries, Table A3.11, p. A32.
70. Ibid., p. 8.
71. Australian Mining Industry Council (AMIC), Mining Taxation and Australian Welfare (Canberra, June 1974), pp. 6-7.
72. Ibid., p. 6.
73. AMIC, Mining Taxation: A Review (Canberra, June 1974), p. 18.
74. AMIC, Mining Taxation and Australian Welfare, pp. 3-4, 7-10, 12.
75. Fitzgerald Report, pp. 45-6, 75, 83.
76. IAC, Report, Petroleum and Mining Industries, p. 2.
77. See, for example, H.G. Raggatt, Mountains of Ore (Lansdowne Press, Melbourne, 1968), p.6.
78. H.G. Raggatt, for instance, felt that one of the most important effects of the concessions was that they prevented delays in the development of Australia's minerals. Raggatt, Mountains of Ore, p. 52.
79. IAC, Report, Petroleum and Mining Industries, p. 2.
80. This and figures relating to other concessions were derived from IAC, Report, Petroleum and Mining Industries, Table 2, p. 54.

Chapter Seven

THE IMPACT OF FOREIGN-FINANCED MINERAL DEVELOPMENT
IN AUSTRALIA

It is particularly difficult to examine the impact
of foreign-financed mineral development in the
Australian case as official and other statistics
usually cover all projects, regardless of ownership.
This must be borne in mind when considering the
impact of mining on, for example, employment,
government revenue and export earnings. However in
considering the economic and other issues raised by
mineral development, the discussion can usually be
made specific to foreign mining investment.

Generation of Government Revenue

Table 7.1 gives details of royalties and income
taxes paid by the mining industry over the period
1965-77. The absolute level of both rose
substantially during these years, with much of the
increase coming after 1974, reflecting rising
royalty rates, changes to income tax legislation
introduced in 1973/4, and the non-availability of
depreciation allowances claimed at an accelerated
rate in earlier years. The Commonwealth also raises
revenue through the coal export levy, introduced by
the Labor government and now being phased out ($90
million in 1977), and through withholding taxes on
interest and dividend payments made abroad ($10
million in 1977).[1] Direct taxation from mining has
accounted for a small but increasing proportion of
the Commonwealth government's total tax revenue, 0.5
per cent in 1966, 2.3 per cent in 1977.[2] The contribution of royalty payments to total state
taxation has been similarly modest; in 1977, only in
Western Australia did they account for more than 10
per cent of total taxation.[3]
   Perhaps the most persistently-expressed fear of
Labor Party spokesmen was that foreign mining

Table 7.1: Royalties and Income Taxes Paid by the Australian Mining Industry, Selected Years 1965-71, and 1972-7 ($000s)

| | 1965 | 1967 | 1969 | 1971 | 1972 | 1973 | 1974 | 1975 | 1976 | 1977 |
|---|---|---|---|---|---|---|---|---|---|---|
| Royalties* | | | | | | | | | | |
| - To States | 22,994 | 23,689 | 24,960 | 60,528 | 59,109 | 62,082 | 78,062 | 141,714 | 146,326 | 186,734 |
| - To Commonwealth | - | 10 | 11 | 5,024 | 7,567 | 7,895 | 10,769 | 12,155 | 13,440 | 13,805 |
| Total | 22,994 | 23,699 | 24,971 | 65,552 | 66,676 | 69,977 | 88,831 | 153,869 | 159,766 | 200,539 |
| Income Tax | 18,776 | 26,854 | 48,473 | 35,659 | 45,632 | 98,683 | 162,771 | 310,470 | 409,474 | n.a. |

* From 1970 onwards, these figures include royalties received under the Petroleum (Submerged Lands) Act 1967-8.
Source: BMR, Australian Mineral Industry Annual Review, various issues.

companies would undervalue mineral exports and so reduce profits generated within Australia, whether by transfer pricing or by failing to exercise their full bargaining powers in their dealings with customers. Opportunities for transfer pricing have certainly existed, as some mining operations, particularly in the aluminium industry, were conducted by subsidiaries of fully-integrated multi-national mining corporations. In fact on a number of occasions the Taxation Commissioner has imputed prices to mineral exports for tax assessment purposes different to those actually received by the Australian subsidiary.[4] The Commissioner's power to act in this way might be thought to indicate that there is little danger of revenue loss, but in practice he has encountered serious difficulties in establishing 'arms length' prices for mineral commodities which are not freely traded on a large scale (for instance bauxite, alumina).[5]

Foreign mining companies, especially iron ore and coal exporters, have apparently sometimes suffered from failing to adopt a more unified approach in dealing with customers, but it is difficult to differentiate in specific cases between the effect of disunity and that of prevailing market conditions for the commodity involved. This difficulty is illustrated by price negotiations between iron ore producers and the Japanese steel industry in 1978. The Japanese adopted a tactic of concentrating their negotiating effort on one producer. They persuaded the Mount Newman Mining Company to accept a significant drop in prices on the contracts being re-negotiated by offering to re-negotiate an older contract, which was not yet due for renewal, in its favour; this effectively gave the Company a $14 million 'kickback' which balanced the lost revenue from the newer contracts. According to press reports, Mount Newman's acceptance of the lower prices undermined the bargaining position of the other producers, forcing them to follow suit and accept those prices, with none of the benefits afforded Mount Newman.[6] However, the willingness of the other companies to accept a drop in prices reflected the current depressed state of international iron ore markets and the consequent absence of alternative market outlets. Thus Mount Newman's 'defection' was only partly responsible for the failure to obtain higher prices.

## Generation of Employment and Wage Payments

The mining industry is highly capital-intensive, and consequently the numbers employed are small in relation to output. In 1973, capital intensity in mining (measured as the book value of fixed tangible assets per person employed) was five times higher than the average for the manufacturing sector.[7] Direct mining employment as defined by the Australian Bureau of Statistics (ABS) increased from 48,589 in 1966 to 66,074 in 1978; in the latter year, such employment occupied 1.4 per cent of the civilian workforce in Australia. However, this figure underestimates the impact of mineral development on employment, for three reasons. First, it represents a narrow interpretation of mining employment, referring only to those engaged in extracting minerals, and excludes employment in mineral processing. Second, Australia produces a high proportion of the goods and services consumed by the mining industry, and consequently indirect employment is substantial. According to the Australian Mining Industry Council the multiplier effect is about 2.75,[8] indicating that between 150,000 and 200,000 Australians, or around 3.5 per cent of the workforce, depend directly or to a large extent on mining (as defined by the ABS) for their jobs. Third, where mining occurs in remote areas almost the entire workforce usually depends on mining and related activity for employment. For example, in November 1976 it was estimated that 87 per cent of the workforce in the Queenstown area of Tasmania were dependent on the Mount Lyell copper mine through direct wages, contracts, purchases, or expenditure of mineworkers' wages.[9]

It is difficult to assess the economic significance of mining employment since general economic conditions in Australia have changed substantially over the period under review, and since significant variations in such conditions occur on a regional basis. In the 1960s Australia experienced virtual full employment and the mining industry cannot therefore be regarded as having generated employment from a national point of view, since the labour it absorbed could have been utilised elsewhere. In this situation the principal impact of mining employment would probably have been on the incomes of the existing workforce, since the additional demand for labour could be expected to have pushed up wages. Two other aspects of this question should be mentioned. First, by 1970 gross product per person employed in mining was higher than in any other sector and twice as high as in manufacturing,[10] indicating that labour employed in mining was utilised more productively than it would have been in most alternative uses. Second, wages in the mining industry have been substantially higher than the national average, by between 18 and 38 per cent over the period 1966-78,[11] allowing the individuals involved to earn more and adding to the multiplier effects created by expenditure of their incomes.

Since the early 1970s unemployment in Australia has increased steadily, and by 1978 stood at between 5 and 6 per cent of the workforce. Mining activity could consequently be seen as generating employment from a national perspective, but this would only hold true if labour was geographically mobile. This is apparently not entirely the case, and labour shortages still exist in certain remote areas in which mining occurs. Thus the economic significance of mining employment cannot be assessed in general terms; it might be quite different, for example, in the case of the Woodlawn lead/zinc mine, which operates near population centres in New South Wales, than for the Agnew nickel project, situated in a remote part of Western Australia.

Creation of Additional Economic Activity through Linkage Development

Backward Lingage. Australia has supplied a substantial and growing proportion of the goods and services utilised by the mining industry. As in the Irish case, an initial lag in linkage development apparently occurred, and a significant proportion of the capital expenditure associated with mineral developments in the late 1950s and early 1960s was expended abroad, particularly on transport, electrical, and mining equipment.[12] However domestic manufacturing capacity was already well developed and as Australian firms became aware of the mining industry's needs they were well placed to take advantage of the available opportunities, especially since the large size of Australia's mining industry allowed them significant economies of scale. For example, in the 1960s dump trucks used in iron ore and bauxite mining were imported, but by the mid 1970s replacement vehicles were being produced in Australia. It is also apparent that in later years at least some foreign mining companies adopted a deliberate policy of favouring domestic producers.[13]

Foreign-Financed Mineral Development in Australia

By the early 1970s, some 85 per cent of capital investment by mining companies was being expended in Australia.[14] Specialised transport and mining equipment was (and is) still being imported, and this will apparently continue because of economies of scale in the producing industries, though some commentators believe that Australia could establish export markets and thus render local production feasible.[15] Currently, mining firms obtain an estimated 88 per cent of their operational supplies and services within Australia.[16]

While a high proportion of the inputs used in mining are supplied locally, total linkage from the industry is comparatively weak. Over the period 1972/3 - 1975/6, for example, each dollar of output required expenditure of only 30 cents on materials, electricity, fuels, payments to mining contractors, repair and maintenance and freight charges; the equivalent figure for the manufacting sector was 64 cents.[17]

Forward Linkage. The degree to which mineral processing occurs in Australia varies from commodity to commodity. In nearly all cases, production for the domestic market is fully integrated; however, a high proportion of Australia's mineral exports are in unprocessed form, particularly exports of iron ore, mineral sands and, to a lesser extent, bauxite.[18] The attention of policy-makers and of the public has focused on the issue of whether Australia's minerals can or should be subject to a greater degree of processing prior to export, and it is on this question that attention is concentrated here.

The desirability of increasing the proportion of processed metals in Australia's mineral exports has caused less controversy than almost any other aspect of mineral policy, at least until very recently. Federal and state governments of all political hues, official inquiries into Australia's economic circumstances, trade union spokesmen and many academic and other commentators have lauded the likely benefits of following this course of action.[19] State governments have attempted to encourage processing by making the granting of mineral rights conditional on the lessee's willingness to construct smelters or refineries within their states, and by offering financial incentives. For example, in 1957 the Queensland government granted a lease to Comalco Limited to mine bauxite deposits at Weipa on condition that the Company would construct a plant in Queensland to convert its output to alumina.[20] The Queensland and federal governments offered financial incentives by applying a lower royalty rate to bauxite which was to be processed locally,[21] while the Western Australian government encouraged the Western Mining Corporation Limited to smelt and refine its output of nickel ore in Western Australia by providing for a lower royalty at the mining stage if processing was undertaken and by offering to ship the Company's nickel products at concessional rates on the State railway system.[22]

Enthusiasm for mineral processing reflects a belief that it greatly increases value added and so government and export revenue, generates additional employment, creates technical 'spin-offs', generally increases the level of industrialisation, and creates opportunities for decentralisation of economic activity and population. This belief results in part from the fact that processing activities have in the past led to development or expansion of other economic activities. For example, smelting of lead and zinc concentrates mined at Broken Hill resulted in the production of sulphuric acid at Broken Hill itself and in Tasmania, with the acid in turn being used in the production of fertilisers at Newcastle. By-product copper, silver, gold, cadmium and antimony was also produced; some of these metals were used in other industrial processes, others were exported. Another by-product, zinc dust, was utilised in gold metallurgy and paint production. Each of these activities created demands for goods and services, and there can be little doubt that, overall, significant additional employment, incomes, and government revenues were generated as a result of processing activity, a fact of particular significance to state governments in pursuit of 'development'.

The second principal basis on which processing has been sought is that it would substantially enhance the value of Australia's mineral exports. A comparison frequently quoted is that one dollar's worth of bauxite can be converted into $100 dollars' worth of aluminium.[23] Though an extreme case, this example does illustrate the potential impact of large-scale processing on Australia's export income, and comparisons of this kind have played a major part in fostering the belief that Australia has become a'quarry' for consuming countries and is consequently deprived of much of the value of its minerals.

A number of important points should be made in

relation to these perceived benefits. First, resources utilised in processing have alternative uses, in some of which they might exercise an even greater impact. For example, creation of employment is currently a major concern of state and federal governments. Mineral processing is extremely capital intensive; in 1973, capital intensity in smelting and refining of bauxite, alumina, copper and nickel was eleven times that in the manufacturing sector as a whole.[24] (Recent investments in processing and current investment proposals have been concentrated in this group, especially bauxite/alumina and nickel.) It is therefore clear that, in seeking to maximise employment, resources might be more usefully employed elsewhere. It could of course be argued that the capital employed in processing is primarily supplied by specialised mining corporations who would be unwilling to invest it in alternative uses, but as against this two points can be made. First, where state governments sacrifice royalties at the mining stage and/or supply infrastructure or services at concessional rates, substantial commitments of public capital can be involved. Second, significant amounts of capital have been raised by processing concerns from general banking and finance sources in Australia, and this capital presumably would be available for alternative uses.[25]

Mineral processing absorbs other resources, of course, particularly energy resources, usually provided in Australia by state authorities in the form of electricity. There is a danger that their desire to encourage mineral processing may lead state governments to undervalue fuel minerals used as feedstock in generating electricity, an issue raised by current plans to substantially expand Australia's alumina smelting capacity. Australia's competitiveness as an energy supplier has obviously been a major influence in attracting investment in this area, but in their attempts to capture projects for _their_ states, state governments may reduce energy costs below the level required to maintain Australia's competitive position. To date, they have apparently kept electricity prices at levels which have not been increased to take account of inflation or of the general escalation in energy costs.[26] Thus the states may be deprived of a significant part of the value of their energy resources, with the predominantly foreign-owned smelting companies reaping the benefit.[27] If this occurs, it will not of course be the result of any innate characteristic of the mineral processing industry, but rather of essentially political factors, and especially of Australia's federal system of government.

Mineral processing is one of the most energy-intensive industrial activities carried on in Australia. For instance, usage of fuels and electricity in processing bauxite/alumina and iron ore is seven times that in manufacturing as a whole.[28] Consequently if the energy charges imposed by the states do not fully reflect the value of feedstock minerals, the loss to government will be on a very substantial scale.

Thus the resources absorbed by mineral processing may have alternative and possibly more efficient uses. It should also be noted that in their alternative uses these resources might still make a significant contribution to Australia's balance of payments, by allowing production of goods for export or by permitting import substitution. On the other hand, mineral processing can certainly bring important economic benefits. In sum, proposals for mineral processing require careful assessment in the light of available alternatives, especially if they are going to be subsidised by the provision of public resources at less than their full value. The current attitudes of state and federal governments show little evidence that this requirement is being fulfilled.

Whatever the justification for investing in mineral processing on grounds of economic efficiency, the fact remains that Australian governments would prefer a higher proportion of mineral exports to be in processed form, and the question arises as to whether the foreign nature of mining investment has prevented this from happening. It is evident that a number of 'objective' economic factors have militated against development of forward linkages in Australia. First, some mineral developments were financed by loans raised on the security of long term contracts with consumers of ores and concentrates; such contracts could not have been secured for finished metal, and it is doubtful if the loans could have been obtained without them. Second, Australia, like other mineral exporters, faces substantial tariff barriers when attempting to sell finished metal in the industrialised countries.[29] Third, substantial infrastructural costs can be incurred if processing is to occur at the mine site, costs which competitors in the industrialised countries would not have to meet. On the other hand, substantial internal transport costs may be incurred if processing facilities are located near urban areas; for instance, the cost of moving bauxite from Weipa to Gladstone is almost equivalent to the cost from Weipa to Europe and Japan.[30] Fourth, market outlets for certain metals, for example aluminium, are cartelised by the existing producers, creating barriers to entry for new producers.

Nevertheless, it has certainly been the case that some foreign mining companies have invested in Australia specifically with the intention of securing sources of ores and concentrates for existing processing facilities abroad. This was the case, for example, with Alsuisse's development of the Gove bauxite deposits, and other investments in the aluminium industry have been similarly motivated. In these instances, foreign companies would clearly be reluctant to undertake processing in Australia if the outcome was to render existing plant redundant. However, horizontal diversification rather than the search for captive sources of supply has apparently motivated most investments outside the aluminium industry[31] and, given that foreign investors have attached little political risk to investment in Australia, it is consequently likely that economic factors of the type discussed above governed most decisions on whether or not to undertake processing locally.

Impact on the Balance of Payments

Mineral exports have made a major and growing contribution to Australia's export income over recent years. At constant prices, the value of mineral exports in 1977 was over seven times that in 1965, while the proportion of export income accounted for by minerals increased from 12 per cent in 1965 to 42 per cent in 1977.[32] Much of this growth has been accounted for by alumina/aluminium, black coal, iron ore and nickel; it is of course in exploitation of these minerals that foreign investment has been concentrated over the last twenty years.

It is impossible to calculate the net balance of payments effect of foreign-owned mining projects as the available information is incomplete, and in any case relates to the mining industry as a whole. Aggregate data indicates that in 1976/7-1977/8,77 per cent of foreign currency inflows (export receipts, equity and loan capital) remained after deduction of dividends, interest and debt repayments payable abroad, and direct imports.[33] This figure omits imports purchased through Australian agents and the import content of domestic purchases. It also overestimates the contribution of foreign-owned companies, as the total flow of dividends abroad should be deducted from their share of export income, and it is evident that the impact of individual foreign-financed projects has been considerably lower than the aggregate figure would suggest. For example, over the period 1961-71 the Queensland bauxite producer, Comalco, received total overseas funds equivalent to $758 million, of which $403 million was paid abroad on direct imports, loan repayments, interest and dividends, indicating a net positive contribution equivalent to 47 per cent of foreign exchange receipts.[34]

Of the identified foreign currency outflows during 1976/7-1977/8, debt repayment accounted for 26 per cent, interest payments for 16 per cent, dividends for 22 per cent, and direct imports for 30 per cent. Government revenue accounted for only 23 per cent of the foreign currency retained in Australia, with wage payments (30 per cent) and purchases of goods and services (47 per cent) accounting for the balance.[35]

The growth in mineral exports has exercised a major impact on Australia's balance of payments. It helped turn the trade deficits which characterised the early and mid-1960s into surpluses, while the investments which permitted expansion of mining activity were accompanied by substantial capital inflows. Over the period 1969/70-1972/3 both these developments, combined with inflows of non-mining capital, resulted in substantial surpluses on the balance of payments; however, import demand did not expand quickly enough to absorb those surpluses, which were accumulated as foreign reserves. The appropriate policy response would have been to revalue the Australian dollar progressively through 1971, but revaluation was delayed until December 1972, by which time speculative capital inflow had exacerbated the situation and Australia was firmly on the path to external readjustment through rapid inflation.[36] The Labor government responded to the balance of payments problem with successive revaluations of the Australian dollar and stringent controls on capital imports, measures which in turn created other difficulties. Export industries (including mining) faced serious problems in maintaining their competitiveness, while companies found it substantially more expensive to raise development capital abroad. It should however be noted that the accumulation of reserves which the Fitzgerald Report interpreted as indicating that Australia possessed sufficient export capacity did not continue after 1973, and since then foreign reserves have been drawn upon to meet rapidly-escalating deficits on invisibles.

Revaluation would have faced the Australian economy with problems of structural readjustment regardless of when it occurred.[37] Thus while the growth of the mining industry has helped to ensure that scarcity of foreign exchange does not impose a constraint on the Australian economy, it has also, in combination with other factors, created other economic problems for Australia.

Provision of Scarce Resources through the Investment Process

Over the last three decades, foreign investors have played a crucial role in providing the enormous amounts of capital required to finance the discovery and exploitation of Australia's minerals. Comprehensive figures are difficult to obtain, but over the period 1964-70 alone total investment by foreign mining concerns was well in excess of $3 billion.[38] In contrast to the situation in other case study countries, such concerns have mobilized significant amounts of domestic capital in Australia (see below).

As mentioned above, Labor politicians generally believed that Australia could have supplied much of the capital required for mineral development without the assistance of foreign companies, claiming that sufficient domestic resources were available and that what was lacking were appropriate mechanisms through which to mobilise those resources, a viewpoint which exercised a major influence on their whole approach to mineral policy. A number of factors appeared to give that viewpoint credence. First, during the 1950s and 1960s savings generated within Australia were sufficient to finance a very high proportion of fixed capital formation; over the period 1957/8-1970/1, this proportion ranged between 82 and 100 per cent.[39] Second, a substantial part of Australian savings were in contractual form, for example in life assurance and pension funds, and therefore apparently presented a suitable basis on which to finance medium and long term investments of the kind required for mineral development. Third, a high proportion of the capital supplied by foreign companies was obtained either from Australian investors or from foreign banks. It has been estimated that, over the period 1964-70, 72 per cent of total investment by foreign mining concerns was financed from these sources,[40] and the question arose as to whether this capital could not be mobilised by Australian companies.

Though the absolute amount of savings generated in Australia was almost sufficient to finance total capital formation, this did not necessarily mean that sufficient capital of the type required to finance mineral development was available. In fact a number of powerful influences operated to reduce the availability of domestic capital to mining concerns. First, government regulations restricted the ability of banks and other financial institutions to allocate their funds freely. For example, life assurance companies, pension funds and savings banks were effectively compelled to invest part of their capital in government securities or in specific activities (for instance house building). Second, institutional investors displayed a strong reluctance to undertake what were perceived as high risk investments in the mining sector, particularly in exploration. This hostility towards mining investment lessened as the mineral industry became more established, and during the 1970s a number of institutional investors committed large amounts of capital to individual projects; however, investment in exploration continued to be shunned.

These two factors meant that, especially during the 1950s and 1960s, private investors would have had to provide a substantial proportion of mining capital, mainly through the stock exchanges. However, funds from this source were limited; for example, over the period 1967-70, total new capital raisings by all firms on Australian stock exchanges were only $2.5 billion,[41] a significant proportion of which would have been subscribed by institutional investors. In addition, private investors also attached significant disadvantages to mining investments. In general, they displayed a preference for investments with short maturity periods, and few 'grass roots' mining projects qualified in this regard. Investors in exploration ventures would of course have to wait even longer for a (more uncertain) return. Some private investors did subscribe to mining stocks during boom periods, but generally their aim was to obtain a quick return rather than undertake a long-term commitment.[42] Some were also prepared to invest in large, established mining concerns which were better able to bear the risks of exploration, but of course this was of little use to small Australian companies attempting to develop mineral discoveries, while the larger companies were predominantly foreign-owned.

Neither would domestic mining concerns necessarily share the access of foreign companies to overseas borrowings. In many cases, the ability of Australian subsidiaries to raise loans abroad depended on the connections of their parent companies with international financial institutions, connections which domestic firms might not possess. So, for example, Alcoa of Australia attempted unsuccessfully in 1966 to raise overseas loans, but with the assistance of its parents, loans were obtained 'from companies and institutions which had contact with Alcoa U.S.'[43] In other cases, availability of foreign loans resulted from the ability of parent companies to themselves supply assured outlets for minerals produced or to negotiate long term contracts with third parties.[44] In general, Australian companies did not in fact enjoy access to foreign funds on an equal basis with their foreign counterparts.[45]

A number of points arise from the above discussion. First, the availability of domestic capital increased significantly as the projects involved reached maturity, but by this stage ownership was usually firmly in foreign hands. In the absence of Australian investment in exploration, valuable mineral deposits came into the possession of foreign companies; where equity was subsequently issued to Australians, it was usually in small lots and at a substantial premium. In other words, ownership of minerals was alienated because few Australian companies were willing or able to act as risk takers, and the fact that domestic captial was available at a later stage had little impact.

This raises a broader question. Even had sufficient domestic capital been available at each stage of the development process, there was a need for appropriate vehicles to mobilise capital and other resources, for a substantial number of companies which could raise risk capital and, where exploration was successful, marshall the financial, technical and marketing skills required to bring major mineral deposits into production. Some such companies did exist in Australia, but were few in number. There were even fewer concerns which were large enough to undertake massive investments in the large-scale mining and handling facilities and supporting infrastructure required to exploit bulk commodities with low unit values (iron ore, coal, bauxite). The Alcoa venture was illustrative in this regard. It involved three of Australia's largest mining companies, North Broken Hill, Broken Hill South and Western Mining Corporation (WMC), yet the funds required for initial development represented 150 per cent of their combined market value in 1961. In contrast, they represented 7 per cent of the market value of Alcoa US in the same year.[46]

Thus if Australian minerals were to be developed without foreign help or through joint ventures between Australian and foreign companies it was essential to devise mechanisms through which domestic capital could be mobilised and to establish vehicles capable of utilising that capital efficiently in exploration and development. The Labor government recognised this fact, as indicated by its plans to expand the role of the AIDC, establish the National Investment Fund and the PMA, and generally increase direct government participation in mineral development. The question remains as to whether these initiatives could be effective in practice; this issue is dealt with below, but one general point should be made at this stage. Australia's savings were certainly not greater than her capital requirements; savings ratios were already very high and were unlikely to increase,[47] and thus if domestic capital was mobilised for mineral development it would very probably have to be withdrawn from other uses, as the opposition claimed.

Foreign investors have played an important role in providing technology and technical expertise, though the significance of their contribution has varied from sector to sector of the industry. Australia has a long history of technical innovation in mining and metallurgy, particularly in association with working of lead, zinc, copper, and gold, but the expertise required to exploit the 'new' minerals discovered after 1945, particularly aluminium and nickel, was not available locally. For example, major bauxite deposits were discovered in Western Australia and Queensland by two Australian-based companies, WMC and the Zinc Corporation, in the 1950s. Though proficient in mining skills, these companies lacked access to the technology and expertise required to design, construct and bring into operation facilities for producing alumina and aluminium.[48] WMC and the Zinc Corporation formed joint ventures with Alcoa and Kaiser Aluminium respectively, two US companies which had for many years played a major role in the world aluminium industry. The latter used their research facilities and their extensive experience to design processing facilities appropriate to the individual bauxite deposits, and supplied skilled expatriate personnel to supervise construction and operation of those facilities. In both cases, expatriate staff numbers declined as Australian personnel were trained in relevant skills. However, specialised expatriate staff continued to be seconded to Australia, partly to facilitate the transfer of newly-acquired expertise from the parent company to the subsidiary or associate, while Australian staff visited the US for the same reason.[49]

It should be noted that in these cases investment in mineral development came about partly as a result of the foreign companies' possession of relevant technolgy.[50] Alcoa US, for instance, refused to make its technology and expertise available to WMC unless it was offered a 51 per cent shareholding in the Australian operation.[51]

This background did not augur well for the Labor government's hopes that foreign technology and expertise could be obtained without direct foreign investment. Those hopes were in fact dealt a severe blow at the Australia-Japan Ministerial meeting at which they were first expressed officially. The Japanese Ministers refused to study the feasibility of establishing a uranium enrichment plant utilising Japanese technology in Australia so long as the Australian government precluded Japanese equity participation in the project.[52] If the government of a country utterly dependent on imported fuels took this attitude towards an energy-oriented development, it was very unlikely that other potential suppliers of technology would accept the Australian proposals.

The issue of the appropriateness of technology employed by foreign investors is enormously complex, because of the variety of mineral and ore types worked in Australia. However, the outstanding feature of most recent developments has been the use of capital-intensive highly-mechanised mining methods, and in most cases use of such methods was unavoidable if bulk commodities with a low unit value were to be exploited efficiently. Labour scarcity and the remote location of many deposits reinforced this requirement.

Creation of External Economies

Because of the very specialised nature of many of

the processes involved, it is unlikely that non-mining firms have benefitted significantly from technology or expertise imported by foreign mining companies. In the aluminium and nickel industries, ownership of the (sometimes secret) techniques is closely guarded, and consequently little 'spill over' to other mining companies occurs; each new venture makes its own arrangements with foreign suppliers of the technology involved. In areas where access to technology is more widely dispersed, the situation has been somewhat different. For instance, techniques for mining and processing of copper, lead, and zinc developed by or supplied to Mount Isa Mines and CRA have been licensed and made available to other firms. However, the high degree of foreign ownership in the mining industry means that relatively few Australian companies exist which might benefit from any such transfer of technology.[53]

Many recent mineral developments have been located in remote and sparsely-populated regions, and have consequently been accompanied by major infrastructural investments. For example, Hamersley's development of its Western Australian iron ore deposits required construction of roads, railways, and port and shore facilities, provision of power and water supplies, and the establishment of four townships. However, much of this infrastructural development has been particular to mining operations, reflecting the general absence of other forms of economic activity in remote regions. For instance in the Pilbara ports and railways are used almost exclusively for shipment of minerals, while townships and community facilities mainly serve mining company employees. In Queensland, railways associated with some mining projects (for example Mount Isa Mines, Utah's coal developments) pass through established pastoral areas, and have facilitated graziers' access to markets and lowered their transport costs.

Mining company spokesmen and state and federal politicians frequently express confidence that mine infrastructure will facilitate the establishment of new industries in Australia.[54] To date, little evidence has emerged which might justify this confidence. Many of the projects involved have not of course been operational for very long, but the experience of well-established mining centres such as Mount Isa and Broken Hill, both of which have failed to establish significant industries autonomous of mining, is not encouraging.

Foreign-Financed Mineral Development in Australia

## Creation of External Diseconomies

Foreign mining companies have absorbed significant amounts of Australian resources, particularly labour, capital, and of course minerals. The labour utilised in mining would until recently have had alternative uses, and this is still the case in certain regions; however, as mentioned above, labour was probably employed more productively in mining than it would have been in its alternative uses.

Absorption of domestic capital raises more complex issues. The sums involved have certainly been substantial; over the period 1964-72 alone, foreign mining concerns raised over $620 million in loans and $670 million in equity from Australian sources.[55] Could this capital have been more profitably employed elsewhere? As regards loan capital, foreign mining companies had to compete with other potential borrowers, and presumably their success in raising funds reflected their ability to pay interest at rates as high or higher than alternative users. However, where Australian lenders attached a lower element of risk to investing in a subsidiary of a multinational corporation, Australian borrowers might be denied loan finance despite their willingness to pay competitive rates of interest. If the lenders' risk assessment was in fact mistaken, a failure to maximise returns on capital would result.

As regards equity capital, a distinction should be drawn between equity subscribed at an early stage of the development process and that subscribed to established mining operations. As mentioned above, there is evidence that returns on Australian equity subscribed to established foreign-owned mining projects have been low, and to date most major equity subscriptions have fallen into this category. As an extreme example, it has been estimated that the AIDC would have earned an after-tax return of about 2 per cent if it had taken up CRA's share subscription offer (see above, p. 169).[56] The willingness of Australian investors to subscribe equity in this manner can be explained by their expectations of receiving rates of return similar to those enjoyed by the original investors. The case of Hamersley Holdings Limited illustrates how mistaken such expectations can be; as of February 1975, Australian shareholders in Hamersley had paid on average 3.67 times as much for each share held as had the principal original shareholder, CRA.[57] In sum, it is apparent that much of the domestic equity capital absorbed by foreign mining concerns might have been employed more profitably elsewhere.

Foreign investors have also used large quantities of non-renewable mineral resources. The issue here is whether mineral development should have proceeded at a slower pace, allowing Australian investors an opportunity to develop the capacity to exploit those resources. How long it would have taken them to develop that capacity is a matter for conjecture, but policy-makers in the 1950s and 1960s had few doubts as to the answer - large-scale domestically-financed exploration and development seemed very far away, while a high value was placed on immediate returns from mineral exploitation, particularly in terms of foreign exchange earnings.

It is unlikely that foreign mining investment has had a major impact on domestic income distribution. Such investment probably has exercised some upward impact on general wage levels, by creating additional demand for labour in a situation of full employment. However, the numbers employed in mining were relatively small (less than 2 per cent of the workforce), and consequently the direct impact is unlikely to have been substantial. For the same reason and because of the remoteness of the projects involved it is also unlikely that a significant 'demonstration effect' occurred.

This analysis permits some conclusions regarding the fears and assumptions underlying Labor policies. Certain foreign mining companies did engage in pricing and marketing practices likely to result in loss of Australian government revenue, and some (though probably few) may have failed to develop opportunities for viable mineral processing projects. Overseas investment in mining certainly did lead to a substantial outflow of incomes in the form of dividends and interest payments. But was it necessary to increase direct Australian participation in the mining industry to avoid these costs?

As regards transfer pricing, the essential problem arose from the difficulty of establishing 'arms length' prices for the commodities being exported. It is not obvious that greater Australian participation could solve this problem at a cost commensurate with the potential gains. If Australian participation were limited to the mining stage, difficulties would probably arise in disposing of mineral products previously traded within vertically-integrated multinational corporations. On the other hand, attempts to refine those products into finished metal in Australia might frequently prove to be uneconomic, for reasons outlined above; since private Australian concerns would certainly be unwilling to subsidise such operations, the Commonwealth would have to do so. Consequently it seems that little could be done other than to apply the Taxation Commissioner's existing powers as rigorously as possible.[58]

There apparently was a need for government involvement in certain cases to ensure that fragmentation and/or competition among mineral producers did not contribute towards under-pricing of mineral exports. This is not to suggest that the Australian government should have or could have cartelised producers and thus obtained monopolistic prices, or that intervention was required in relation to all minerals at all times, but rather that selective initiatives might be needed to ensure that producers obtained full current market prices for their exports.

Was it necessary to increase direct Australian participation in mineral development in order to prevent underpricing, or did the application of export controls offer sufficient protection? Foreign firms might be less willing to voluntarily cooperate in negotiating with customers than would Australian companies, either because of relevant home country legislation (for instance US anti-trust laws) or because of equity ties with the customers involved. However, there is no evidence that foreign companies have been able to avoid government-imposed discipline. Thus increased Australian participation might reduce the need for government to intervene, but would hardly enhance its ability to prevent under-pricing.

In those instances where failure to process minerals was due to the requirement of overseas companies for guaranteed sources of raw materials, increased domestic participation might result in greater local processing. However, a similar result could possibly be achieved by favouring foreign companies without such a requirement in the granting of mining leases, leaving domestic resources free for alternative uses.

Greater Australian participation in mineral development clearly was essential if the flow of incomes to overseas suppliers of capital was to cease. The important question in this instance is whether the Labor government was justified in its assumption that such participation could be achieved without significant decline in the efficiency or scale of mining operations or in overall levels of investment. It is evident that its expectations regarding the willingness of foreign mining companies to supply their resources (technology and technical expertise, entrepreneurship and risk capital) without being offered equity participation were largely mistaken. This misperception resulted from two factors. First, the 'oil crisis' of 1973/4 and the commodity boom which accompanied it distorted perceptions as to the true long-term demand/supply situation for energy minerals other than petroleum/natural gas and for non-fuel minerals, hiding the fact that alternative sources of supply would be available for most of these minerals for the foreseeable future, and that foreign investors would consequently not be prepared to pay a high premium for access to guaranteed supplies from Australia. Second, foreign investors in fact generally equated 'security of supply' with the ownership and control conferred by direct equity participation, a point recognised by the Labor government towards the end of its term in office.

Thus additional domestic resources would have to be mobilised if Australian participation was to increase. As mentioned already, domestic capital mobilised for mining investment would probably have been withdrawn from other uses. If investment levels were not to decline, Australia would have to finance resource development with capital borrowed overseas, which would mean that interest payments would continue to flow abroad. In fact, this was the course the Labor government apparently intended to take. It is ironic that its attempts to do so brought about a constitutional crisis, its removal from office, and the abandonment of its schemes to increase Australian participation in mineral development.[59]

There was also a need for Australian corporate vehicles which could act as risk-takers, apply mining capital efficiently, and marshall the other resources required for mineral development. Because of the shortness of Labor's term in office and of the legislative delays caused by the Senate and by legal challenges, it is a matter for conjecture whether the necessary entrepreneurial and other skills could have been provided by a combination of private Australian firms (assisted by the AIDC and the PMA) and government mining corporations. It seems likely that a substantial period of time would have elapsed before these concerns could develop the technical expertise and the financial strength to undertake large-scale developments alone, and conse-

quently foreign investors would probably have
continued to play a major role. In addition, alien-
ation of mineral ownership had already occurred to a
considerable degree. For both these reasons, the
level of foreign ownership in the mineral industry
was likely to remain high for some time to come.

It was consequently important that taxation
provisions applying to the industry should appropriate a substantial proportion of any economic rents
arising from mining.[60] This was so for another
reason. The analysis in this Chapter indicates that
the Labor Party was justified in its pessimism
regarding the likelihood of substantial direct
economic benefits arising from mining projects.
Mining was highly capital-intensive and so had
little employment-generating potential, backward
linkage was comparatively weak, mineral processing
might not create substantial benefits relative to
the alternative uses of the resources involved, it
was unlikely that mining infrastructure would give
rise to significant additional non-mining activity,
and there was little 'spillover' of technology from
the industry. Thus taxation of mining profits would
constitute an important element in any benefits
which Australia would receive from exploitation of
its minerals, and Labor's initiatives aimed at
increasing tax payments by mining companies were
therefore appropriate.

Notes

1. AMIC, Minerals Industry Survey 1977 and 1978
(Canberra, 1979), Table 4.5, p.37.
2. Table 7.1 and ABS, Public Authority Finance,
Federal Authorities, various issues.
3. ABS, Public Authority Finance, State and Local
Authorities, various issues.
4. Commonwealth Treasury, Overseas Investment in
Australia, p.53.
5. Ibid.
6. Mining Journal, 8 September 1978, p.177.
7. IAC, Annual Report, 1973/4, Table 4.1.1.
8. AMIC, Mining Review, November-December 1977, p.2.
9. IAC, Interim Report, Copper Ores and Concentrates, 30 September 1977 (Government Printer,
Canberra, 1977) p.21.
10. W.D. Scott and Co. Pty. Ltd., 'The Contribution
of the Minerals Industry to Australian Welfare:
Report to Conzinc Riotinto of Australia Ltd.',
unpublished, Sydney, July 1974, p.2/13.
11. Derived from ABS, Average Weekly Earnings,

various issues, and Mining Establishments: Details of Operations, various issues.
12. For example in implementing its expansion programme over the period 1958-68 Mount Isa Mines purchased electricity generating, ventilation, milling, and transport equipment in the US, the UK, and West Germany.
13. Senate Select Committee on Foreign Ownership and Control, Official Hansard Report, 20 July 1972, p.659, 7 September 1972, p.1306.
14. This figure was quoted by CRA in its submission to the Senate Select Committee on Foreign Ownership and Control, and accords with other published estimates.
15. Queensland Government Mining Journal, October 1979, p.514.
16. AMIC, Minerals Industry Survey 1977 and 1978, p.52.
17. Derived from ABS, Mining Establishments, Details of Operations, 1975/6, Tables 4 and 6, Manufacturing Establishments, Details of Operations by Industry Class, various issues, Tables 5 and 7.
18. For a comparison of the degree to which processing of various minerals occurs in Australia, see BMR, Australian Mineral Industry Review 1977, Table 12, p.13.
19. See, for example, Age, 20 March 1973; Mining Journal, 9 December 1977, p.478; Queensland Government Mining Journal, July 1974, pp.233-4.
20. Queensland, Commonwealth Aluminium Corporation Pty. Limited Agreement Act, 1957, Eliz. 2, No.29, Clause 7.
21. Northern Territory of Australia, Mining (Gove Peninsula Nabalco Agreement) Ordinance 1968, No.15 of 1968, First Schedule, 3, (b).
22. Western Australia, Nickel Refinery (Western Mining Corporation Limited) Act, 1968, 17 Eliz. 2, No.24, Clause 3 (1) and (2), Clause 5 (a) and (b), Clause 9 (1).
23. See, for example, 'Is the aluminium worth the energy?', National Times, 16 December 1978.
24. IAC, Annual Report, 1973/4, Table 4.1.2, p. 81.
25. For example in 1971 the Queensland alumina producer, Queensland Alumina Limited, had invested $35.5 million provided by Australian institutional lenders.
26. 'Is the aluminium worth the energy?', National Times, 16 December 1978.
27. This assumes that the cost of energy to Australian producers would not significantly affect international prices for alumina/aluminium.

28. Derived from ABS, Mining Establishments, Details of Operations, 1975/6, Tables 5 and 8.
29. A number of foreign-owned Australian producers have been forced to purchase or build processing facilities abroad to overcome this problem.
30. CRA, 'Submission to the Industries Assistance Commission, Inquiry on Petroleum and Mining Industries', unpublished, May 1975, p.413.
31. R.B. McKern, Multinational Enterprise and Natural Resources (McGraw-Hill, Sydney, 1976) pp.154-5.
32. Derived from BMR, Australian Mineral Industry Annual Review, various issues.
33. AMIC, Minerals Industry Survey 1977 and 1978, Table 6.1, p.53.
34. Derived from Senate Select Committee on Foreign Ownership and Control, Official Hansard Report, 20 July 1972, Appendix 11, p.781.
35. Derived from AMIC, Minerals Industry Survey 1977 and 1978, Table 6.1, p.53, Table 4.5, p.37.
36. B. Smith, 'Australia's Minerals Production and Trade: Case Study of a Resource-Rich Developed Country', in Krause and Patrick, Mineral Resources in the Pacific Area, pp.259-60.
37. For a discussion of this point, see R. Gregory, 'Some Implications of Growth in the Minerals Sector', Australian Journal of Agricultural Economics, Vol. 20, No. 2.
38. McKern, Multinational Enterprise, Table 8.2, p.162.
39. Commonwealth Treasury, Overseas Investment in Australia, Table 26, p.39.
40. Derived from McKern, Multinational Enterprise, Table 8.2, p.162.
41. Fitzgibbons, Economic Contribution of the Mineral Sector, p.3.
42. McKern, Multinational Enterprise, p.167.
43. Senate Select Committee on Foreign Ownership and Control, Official Hansard Report, 20 July 1972, p.647.
44. This was the case, for instance, in relation to foreign loans obtained by Queensland Alumina Limited to finance construction of the Gladstone alumina refinery.
45. Senate Select Committee on Foreign Ownership and Control, Official Hansard Report, 29 June 1972, p.179.
46. Ibid.,20 July 1972, p.647.
47. The government could of course attempt to enforce a higher level of savings by increasing taxation, but any such action would probably be

counteracted by a decline in voluntary savings.
48. McKern, Multinational Enterprise, pp.178-9.
49. Senate Select Committee on Foreign Ownership and Control, Official Hansard Report, 20 July 1972, pp.647-9, 764-5.
50. It also resulted from their access to sources of finance and to market outlets for the output of Australian bauxite and alumina.
51. Senate Select Committee on Foreign Ownership and Control, Official Hansard Report, 20 July 1972, P.645.
52. Age, 31 October 1973.
53. Johns in Sinden, The Natural Resources of Australia, p.297.
54. See, for example, Senate Select Committee on Foreign Ownership and Control, Official Hansard Report, 20 July 1972, p.792; Raggatt, Mountains of Ore, p.75.
55. Derived from McKern, Multinational Enterprise, Tables 8.2, 8.3, p.162.
56. Age, 14 April 1973.
57. Derived from CPD, Vol. 93, 19 February 1975, p.508, Table II.
58. The Australian government could have attempted to relate export prices directly to the value of finished metal. Attempts by the Jamaican government to adopt this approach in relation to bauxite exports led the companies to reduce output in Jamaica, compensating with increases elsewhere; thus a united front would have to be formed among mineral producing countries before such an approach could succeed.
59. National Times, 29 August - 3 September 1977.
60. Failure by government to appropriate economic rents in an Australian-owned industry might not be cause for concern, since the incomes involved might be taxed at some other point or be reinvested in productive activity elsewhere in the economy.

Chapter Eight

MINERAL DEVELOPMENT IN PAPUA NEW GUINEA: THE
COLONIAL PERIOD, 1964-1972

Papua New Guinea consists of the eastern portion of
the island of New Guinea and some 600 smaller
islands, a total land area of 463,000 square kilo-
metres. Its topography is rugged; much of the
country is covered by steep mountains, while some 70
per cent of its land area is forested. Many of its
rivers are unnavigable, there are no railways, and
the capital, Port Moresby, is not connected to any
other major urban centre by road. Because of its
formidable geography, the country is culturally
fragmented; over 700 local languages have been identified, while a wide variety of social structures
exist, though most share two characteristics - a
social security system through which members of
kinship groups assist and support one another, and a
system of land tenure in which ownership is a community right granted to individuals or families by
agreement among owning groups.

Permanent contact with Europeans was not established until after 1870, when missionary, labour-
recruiting, trading and plantation activities commenced. In 1884 Britain and Germany divided the area
which now constitutes Papua New Guinea, Britain
taking control of the southern mainland and the
southern islands (Papua), Germany taking the
northern mainland and its neighbouring islands (New
Guinea). After the First World War, both areas were
administered by Australia, separately until the
Second World War and jointly from Port Moresby (as
the Territory of Papua and New Guinea) thereafter.

Until the early 1960s Australian policy-makers
assumed that Australia would administer the Territory for a long time to come, and little was done to
encourage political participation by indigenous
people. However, political development proceeded
very rapidly from 1970 onwards. In 1973 Papua New

Guinea was granted self-government and in 1975 it became independent. A Westminster-style parliamentary system was established, and this has since been modified by the creation of a system of provincial governments.

Papua New Guinea's economy is predominantly agricultural; subsistence agriculture (increasingly augmented by cultivation of export crops) engages from 60 to 70 per cent of its 3.0 million inhabitants, while over 50 per cent of its 200,000 wage earners are engaged in primary industry. Only 7.2 per cent are occupied in industrial activity.[1] Until very recently, Papua New Guinea's external transactions have been balanced by, and much of its public expenditure financed by, Australian government grants and expenditures. This situation has changed to some extent over the last ten years, primarily because of the development of a handful of foreign-financed export-oriented natural resource projects, the most important of which is the copper/gold mine on Bougainville Island. In this Chapter and the next the development of the Bougainville mine is described, and its impact on the economy, society and politics of Papua New Guinea is analysed.

Historical Introduction

In 1961 an Australian government geologist, J.E. Thompson, compiled a report which concluded that the Panguna gold-copper prospect on Bougainville Island warranted investigation. At that time the London-based Rio Tinto Zinc Corporation (RTZ), through its subsidiary CRA, was prospecting for low-grade copper deposits in Eastern Australia. A company geologist, Ken Phillips, persuaded CRA that New Guinea offered a more favourable geological climate in which to search for such ore bodies. CRA decided to investigate the Panguna prospect, and in December 1963 it obtained a Special Prospecting Authority for the area.

Bougainville is the principal island of the northern group of the Solomon Islands. The steep Crown Prince mountain range dominates the Island; flat land is scarce, though Bougainville's volcanic soil is extremely fertile. The Island's political status was determined in 1898 when Britain gave Bougainville to Germany in exchange for the rest of the Solomon Islands, whose inhabitants shared strong racial and cultural links with the Bougainvillians, and for the Tongan group. In 1914 Australia took

216

over Bougainville from Germany; Australian rule was characterised by an attitude of laissez-faire, if not by outright neglect, though individual officials tried hard to assist the local people. Complaints of Administration neglect emanated from every province in the Territory during the 1960s, but Bougainville apparently received scant attention even in relative terms.[2] The Administration may have felt that the Island was comparatively well-endowed in natural resources and also in terms of social services (provided mainly by missionary groups), and consequently should yield priority to less fortunate regions. However, this circumstance did not detract from the sense of neglect felt by Bougainvillians. The Islanders also apparently suffered at the hands of individual Europeans, enduring 'specific acts of discrimination including occasional physical violence'.[3]

Little accurate information exists regarding the economic or social character of Bougainville in the decades prior to 1963.[4] Commercial agriculture had commenced in the 1880s, when Australian and European interests established coconut plantations in coastal areas. Relatively few Bougainvillians accepted the wage employment offered by the planters, preferring to rely on subsistence production, and plantation labour was drawn from Highland and other districts, particularly in the post-war period.

By the early 1960s, about 75 per cent of Bougainville's 72,000 indigenous inhabitants were still engaged in agriculture. Output of cocoa and copra was substantial by this stage.[5] Large-scale production was dominated by expatriates, but many local people were producing cash crops to supplement subsistence income. It has, however, been estimated that 41 per cent of the indigenous workforce were still entirely engaged in subsistence production in 1966.[6] Aggregate figures distort the picture somewhat; it is likely that villages located away from flat, fertile coastal areas were considerably more dependent on subsistence farming.

In 1964, Bougainville was an island whose people harboured a sense of resentment, developed over many decades, against the Australian Administration and against Europeans in general. The Bougainvillians were also a people whose whole mode of existence depended on access to and use of agricultural land.

The Early Exploration Period, 1964-6

In January 1964, Phillips travelled to Kieta to prepare for a geological survey of the Panguna prospect,[7] located at a height of some 3,500 feet in dense tropical forest, and in mid April 1964 set out for Panguna with three assistants. Initially, no objections were made to their presence. In early May a group of local people visited the site to find out what the geologists were doing. A lengthy discussion ensued during which the villagers questioned CRA's right to prospect on their land, but as far as Phillips was concerned they approved the presence of the exploration team. Whatever the case, one group of local people, the Guavas, soon expressed opposition to CRA's activities, and demanded that the geologists leave. Subsequently Max Denehy, the Assistant District Commissioner, pointed out to Phillips that CRA was not employing any Guavas; he called village leaders to Kieta and explained CRA's plans in detail, and Phillips undertook to employ Guavas at the Panguna site. This apparently placated the Guavas, as exploration work was not interrupted for several months thereafter. By September 1964, Phillips was convinced that a major copper deposit existed at Panguna.

In the meantime, CRA sought exclusive prospecting licences for the Panguna area, and a Warden's Court was held in Kieta in January 1965 to consider its application. The Guava people objected to the granting of the licences, and questioned the Court's right to issue them. The Mining Warden explained that existing legislation vested ownership of all minerals in the Administration and thus conferred that right on the Court, but the Guavas denied the validity of the Administration's claim. Despite their objections, CRA's application was successful.

In January CRA also received its first really favourable drilling results; CRA and the Administration's Mines Division now knew that there was a strong probability of a mine eventuating in time.[8]

From January 1965 onwards, Phillips spent much of his time trying to convince local people that the Panguna development would be in their interests, a difficult task since he was unable to say what financial benefits development might bring. The Mining Ordinance then in force did not provide for payment to landowners of occupation fees or of compensation for minerals removed. Phillips and Denehy were concerned with the implications of this situation; they were facing more and more questions as to what local people would gain from CRA's activities, and many villagers were now openly expressing dissatisfaction with the Company's presence.

In mid 1965 Phillips and Denehy submitted a document on the matter to the Administration in Port Moresby. This discussed the problems of the Company and of the local people, offered advice as to how conflict between the two could be avoided, and emphasised the need to promise monetary compensation to local people for minerals removed from their land. The unusual step of submitting such a document was taken because, as Phillips explained, 'we were quite sure that [Port] Moresby had no real appreciation of the situation and Denehy was entirely frustrated in trying to get anything through normal channels'.[9] More immediately, it was prompted by a number of incidents involving CRA personnel and local people. In July 1965 Phillips was absent from Bougainville. His replacement, a senior exploration man from Melbourne, pushed the exploration effort ahead vigorously despite warnings from Denehy and Phillips, and sent a party into Mainoki village without consulting the villagers. The party was expelled and the villagers refused to permit CRA personnel back on to their land.

The most serious incident apparently resulted from a misunderstanding. The Company wished to prospect in an area between Panguna and Kupei, and it consulted the Kupei and Guava people who, it believed, held customary rights in the area. As far as CRA understood, there was no objection. A hunting party from Musinau, a village which, unknown to CRA, held hunting rights in the area, came upon the exploration camp and pulled it down. Police were called in and a number of Musinau men were imprisoned. Following this incident, a substantial body of policemen were brought into the prospecting area. The majority were soon withdrawn but this type of exercise was precisely what Phillips and Denehy had been working to avoid.

Relations between local people and the authorities deteriorated rapidly from this time onwards, and in Phillips' view the visit of the Minister for External Territories to Bougainville in mid 1966 was 'the major milestone' in this process. Mr Barnes was briefed for about half an hour on his arrival at Port Moresby by the Assistant Administrator, Henderson. He flew to Kieta and without consulting Denehy or other individuals on the spot spoke to meetings of local peple. In effect, Barnes told these people that there was nothing in the Panguna development specifically for them, but rather that the project would benefit Papua New Guinea as a whole. Phillips and Denehy were thunderstruck; a new Mining Ordinance was then being prepared and they had conveyed their optimism regarding its likely provisions to the local people. In the wake of the Minister's visit, more and more local people refused CRA employees access to their land. The Administration responded by adopting a 'hard line' policy. Denehy, who was convinced of the need for patience and compromise, was transferred from Bougainville. Additional police were drafted in and a patrol post established at Panguna. By the following year the situation was, according to Phillips, 'appalling'.

In June 1966, amendments to the Mining Ordinance were introduced in the House of Assembly, a legislative body made up of elected representative and of 'official' members appointed by the Administration. The amendments reasserted the principle of Administration ownership of minerals, a principle which Administration spokesmen defended on the basis that mineral exploitation should benefit all Papua New Guineans, not merely those fortunate enough to own land rich in minerals. They pointed to the actions of other national governments in legislating for state ownership as evidence of the concept's validity, and stressed that every dollar of mineral revenue would be needed by Port Moresby to finance the economic development of the Territory as a whole.[10]

The Acts provided for payment of a minimum occupation fee of $1 per acre per annum to landowners affected by prospecting or mining activity.[11] Persons wishing to enter and prospect on private land would have to obtain a permit from an open Warden's Court, thus allowing 'full opportunity for public interest and private interest to be represented'.[12]

The member for South Bougainville, Paul Lapun, was entirely dissatisfied with these provisions. The Acts provided for the imposition of royalties on mineral production, and Lapun tabled an amendment providing that 20 per cent of those royalties would be paid to the landowners involved. He pointed out that, according to local custom, mineral rights belonged to the landowner. Mining would deprive local people of their livelihood as it would permanently damage their land, and substantial compensation was therefore required. Lapun warned

'if the people in Kieta are dissatisfied...trouble
could arise.... If the landowners receive a share of
the potential benefits then they will welcome the
company'.[13]

Lapun's amendment was strongly opposed by the
Administration, which argued that the concept of
state ownership would be undermined by payment of
royalties to landowners. In any case, it was stated,
payment of compensation and occupation fees would
ensure that landowners were not disadvantaged, while
islanders would benefit directly from developments
associated with the mine.[14] Lapun reduced his
demands to 5 per cent of royalties but the Administration stood firm and his amendment was defeated.
He re-introduced it in November 1966, by which time
trouble had indeed arisen on Bougainville. The
Administration again opposed the amendment, but a
number of members were now convinced that some
initiative should be taken. Lapun assured the House
the prospect of obtaining a share of royalties would
satisfy his constituents, and the amendment was
passed.[15]

But Lapun was mistaken - his efforts did little
to improve the situation on Bougainville, where
conflict between the local people and CRA continued.
In fact many villagers refused to accept the
payments due to them lest acceptance be interpreted
as indicating their willingness to alienate their
land.

In the following years, the hostility of local
people would wax and wane; most have now fully
accepted the benefits resulting from mineral development, though many still resent the presence of the
Company and its employees. Certain villages have
consistently refused to have anything to do with the
mining project. The events of 1964-7 were crucial in
determining the attitude of local people to the
Panguna development. What caused the confrontation
and conflict which characterised CRA's early involvement? Could that confrontation have been avoided
or mitigated?

The Causes of Conflict

A certain degree of conflict was probably inevitable, given the nature of CRA's activities and of
Bougainville's economy and society. The variety of
languages spoken on the Island combined with the
very different cultural backgrounds of the individuals involved certainly made it likely that
misunderstandings would arise and that mistakes

would be made. Nevertheless, conflict need not have been as serious nor as extensive as it was.

First, some of CRA's mistakes could have been avoided if the advice of officials 'on the ground' had been heeded. The action of a senior company official in sending geologists into Mainoki village (see above) is a case in point. Second, and more importantly, CRA and the Administration could have done considerably more to try and understand the character of the local people, and to act in a way designed to minimise their resentment and allay their fears. This section outlines and seeks to explain the failure of the Company and the Australian authorities in this regard.

It is apparent that, at least over the period 1964-7, neither CRA nor the Administration made any deliberate attempt to familiarise themselves with the social, cultural, economic or political circumstances of the people with whom they were dealing, though both knew from January 1965 onwards that a mine might well eventuate on Bougainville.[16] Individuals like Phillips and Denehy learned much from personal experience, but their superiors made little use of this knowledge, a fact especially evident in relation to the whole issue of land and land use.

It is difficult to any outsider to grasp fully the meaning of land to a Bougainvillian villager, but the following passage offers some insight.

> Land is our life. Land is our physical life –
> food and sustenance. Land is our social life;
> it is marriage; it is our only world. When you
> [the Administration] take our land, you cut
> away the very heart of our existence. We have
> little or no experience of social survival
> detached from the land. For us to be completely
> landless is a nightmare which no dollar in the
> pocket or dollar in the bank will allay; we are
> a threatened people.[17]

In general, officials showed little appreciation of the importance of land, as indicated by the Administration's failure to offer villagers more than a small yearly cash payment in return for land which, if mined, would be permanently lost to them. In addition, they failed to take into account the nature of land ownership and usage on Bougainville.

Because of the system of ownership and inheritance practised by villagers, numerous individuals other than the current occupier might hold an interest (or potential interest) in any one piece of land. A cash payment might satisfy the occupier, but it offered nothing to other interested individuals. An important point as regards land usage was that its various aspects might be treated quite separately. Land could be built on, planted, hunted on, walked through, or used as a food collecting area, but rights to carry out certain of these activities might belong to individuals or groups other than the occupier or occupiers. Failure to realise this led directly to conflict (for example in the case of the Musinau hunting party), and also generated resentment among those who did not receive occupation fees as compensation for their loss of rights.

Administration officials also showed little awareness of the political feelings of Bougainvillians. They constantly appealed to the islanders for co-operation on patriotic grounds, and resisted Bougainvillian demands for a share of royalties by claiming that mining revenue should benefit Papua New Guinea as a whole.[18] The concern of Australian officials with nation-building is understandable, but in calling on a patriotic sentiment they entirely ignored the political realities: such a sentiment simply did not exist. The focus of loyalty for most Bougainvillians did not extend beyond their clan or district. They certainly had no loyalty to 'Papua New Guinea', and felt that revenue generated by their cocoa and copra industries was already being used for the benefit of other Papua New Guineans, while they themselves received little or nothing from Port Moresby. Bougainvillians simply could not understand why they should make sacrifices so that others might reap the benefits.

This lack of understanding on the part of Administration officials explains why they were so frequently at cross purposes with Bougainvillians, and why they often became frustrated in their attempts to explain and justify their actions. As one official commented:

> No amount of talking and reasoning on any subject whatsoever, will alter their thinking and therein lies the basic problem. Once a decision is made by the people nothing no matter how sound, sane, and logical will be listened to, and unsound, insane and illogical reasoning on their part will be used to defend their decision. [Emphasis in original.][19]

Australian Administrators failed to understand that their actions might be sound, sane and logical within their own frame of reference, but might appear unsound, insane and illogical within the autonomous cultural, social and political structures employed by Bougainvillians to assess initiatives taken by outsiders. The Administration need not have allowed its decisions to be dictated by the islanders, but it should have accepted their reactions to its policies as legitimate expressions of anger, self-defence or fear and, where possible, amended those policies accordingly. To some extent what was involved was a question of attitude. Had the Administration's attitude been different, CRA's activities would have generated considerably less resentment.

Nevertheless, the crunch would have come. Some local people simply wanted CRA to 'wrap up its things and go home', while the Administration was determined not to allow a small group of villagers to obstruct a major mineral development. However though Bougainvillians would inevitably have had to face the fact that mining would occur, and that their lives would be radically changed, the trauma of facing such change could have been mitigated. This is well illustrated by the issue of compensation and royalty payments. Throughout the House of Assembly debates on the royalty issue, Administration spokesmen claimed that Bougainville people were simply anti-mining, that they wanted CRA to leave, and that offering them a share of royalties would have no effect on their attitudes.[20] Yet Phillips and Denehy, the two individuals who had the most direct contact with local people over the preceding years, believed that it was vitally important to offer concrete financial incentives to the villagers[21]. Clearly they felt that though a 'dollar in the bank' might not compensate a villager for his land, it would make his loss more acceptable.

Apart from the royalty question, the Administration's general approach was unsuitable to circumstances on Bougainville. Denehy and Phillips were in favour of compromise, of patient endeavour designed to fully inform people of developments and hopefully to win their approval. They opposed the use of force in response to hostile reaction by local people.[22] The Administration rejected this approach and pushed for vigorous action, meeting resistance with coercion, a strategy implemented under Denehy's successor.

It is obvious that the Administration was unwilling rather than unable to develop a policy suitable to the situation on Bougainville. For example, officials in Port Moresby must have been aware of the sensitivity of the land issue. After all, the Administration had been dealing with indigenous systems of land ownership for fifty years. It received ample advice from Denehy and Phillips on the royalty issue and generally on its dealings with local people, but that advice was ignored.

What explains the Administration's behaviour? To provide an answer, it is necessary to understand the broader context within which policy towards the Bougainville project was formulated. In 1962, a visiting United Nations Mission had criticised the Administration for failing to formulate plans for the economic development of the Territory, a criticism later repeated in a number of United Nations General Assembly resolutions. The Mission also stated there was an urgent need to rapidly increase expenditure on education for Papua New Guineans.[23] In 1963 the Australian government invited the World Bank to draw up an economic development plan for the Territory. The Bank recommended a policy aimed at achieving rapid growth in the monetary sector of the economy, and argued that substantial increases in Australian financial support would be required if this policy were to succeed. Again, the urgent need for increased educational expenditure was stressed.[24]

Thus by 1964 the Administration was anxiously searching for ways in which to stimulate growth in the monetary sector of the economy, and for sources of internal revenue which would allow public expenditures to be maintained, and Australian subsidisation to decline, with the advent of independence. Given Papua New Guinea's economic circumstances, large-scale mineral development offered perhaps the best prospect of achieving both. The discovery of the Panguna prospect was consequently a heartening development for the Administration.

CRA, on its part, was anxious that prospecting be concluded quickly, and it apparently exerted pressure on the Administration to ensure that prospecting teams were not withdrawn when faced with local opposition, lest they be prevented from re-entering the area.[25] CRA's need for haste continued after detailed exploration and project evaluation commenced. It regarded the Panguna prospect as a marginal one whose viability might depend on 'catching' a predicted boom in copper prices. The Administration accepted CRA's viewpoint and believed that if delays occurred it might lose a development which it very much wanted to proceed.[26] The approach advocated by Phillips and Denehy would take time, as indeed would any detailed analysis of the situation on Bougainville. As CRA had convinced the Administration that time was precisely what it did not have, a different strategy was required.

The royalty issue was a different matter. The question of time was not involved, and indeed a concession in this area might well have saved time as local opposition to CRA might have been mitigated. As mentioned above, the Administration claimed to be defending the principle of state ownership of minerals. If that principle was breached, it argued, individual landowners could obstruct development and claim part of mineral revenues.[27] But by 1966 it was apparent that individuals could cause obstruction without possessing mineral rights, and it was clear to officials on Bougainville that financial incentives would have to be offered if local opposition was to be overcome.

This raises the broader question of whether legal concepts considered appropriate in other countries were suitable to local circumstances. Bougainvillians simply could not understand why minerals under their land should belong to the Administration, and given the enormous importance they attached to land it would seem reasonable that Bougainvillian landowners should be entitled to more generous compensation than their counterparts in Western societies. But the Administration, anxiously seeking sources of revenue, was determined to retain all royalty payments for its own purposes, and it was the question of who would receive royalties, rather than any consideration of principle, which was the real issue.[28]

The Administration's economic policies are analysed below, but whatever the economic rationale for its actions, they were interpreted by Bougainvillians as indicating its determination to protect its own interests or those of CRA. When the Company wanted an exploration licence, it was granted.[29] If local people expelled CRA personnel from their land, police were brought in to ensure that prospecting continued. If villagers acted to protect their traditional rights, they were imprisoned. Bougainvillians felt that the Administration was deliberately sacrificing their interests.[30] It was this feeling of being abandoned, at a time when powerful and only partly understood forces were radically changing their lives, which lay behind much of the resentment and anger felt by local people. The

Administration could certainly have done much to allay that feeling without sacrificing the Panguna project. Its failure to do so was to have a long and sometimes bitter legacy.

By late 1966, exploration work at Panguna had reached an advanced stage. Eventually, ore reserves of 900 million tons would be established, grading on average 0.48 per cent copper and 0.36 dwt. per ton gold, and also containing silver in recoverable quantities. As its initial exploration programme neared completion, CRA faced the prospect of investing substantial sums in project evaluation and possibly project development. The Company was therefore anxious to negotiate a contract with the Administration which would spell out the terms under which any development would take place. As the position stood in 1965, CRA would automatically be entitled to a mining lease if and when its deposit was proved commercially viable, but the Mining Ordinance did not specify the terms on which a lease would be granted. The terms of a Special Mining Lease for the Panguna prospect were negotiated over a twelve-month period to June 1967, when an agreement was concluded between CRA and the Administration. Before its terms are outlined, the policy aims and priorities of the two parties will be discussed briefly.

The Administration had two principal concerns, to guarantee a major source of internally-generated public revenue for the future, and to secure an immediate source of export income and of stimulation to the monetary sector of the economy. Administration officials did not believe that internally-generated revenue was needed urgently, since Australian subsidisation was guaranteed in the short term and since the Papua New Guinea economy could not in their view efficiently absorb additional large-scale public expenditures. However, as Papua New Guinea approached political independence (a prospect not likely to eventuate before 1980, it was then felt), and as the economy expanded and became capable of absorbing resources at a greater rate, the need for domestic sources of revenue would increase.

The Administration was anxious to secure a major source of export income as quickly as possible, since the Territory's external balance of trade was permanently and substantially in deficit.[31] It also hoped that the establishment of a major mineral project would act as a catalyst to further private investment, by providing a demand

for goods and services and by generating interest
and confidence in Papua New Guinea as a location for
private foreign investment. The Administration was
therefore prepared to forego revenue in the short
term in order to ensure that the Panguna development
would proceed and thus contribute to the implementation of its broader economic and financial
strategy.

CRA wished, of course, to operate as profitably
as possible at Panguna, and more particularly to
achieve a high level of profits during the early
years of mine life. The Company's preference in this
regard reflected, in part, the application of a
discount rate to future earnings, but it also
reflected CRA's capital requirements. Substantial
loan finance would be needed to help fund a development at Panguna. The prospective lenders would
attach a high risk factor to investment in the
project, partly because of the political uncertainty
they would associate with investment in a developing
country, partly because of economic and technical
factors associated with exploitation of the Panguna
orebody.[32] CRA would have to persuade these lenders
that its cash flow would be sufficiently large to
allow rapid repayment of its borrowings, thus
ensuring that their funds would be at risk for the
shortest time possible. The Company would therefore
be prepared to accept high taxation rates at a later
date in return for concessional rates during the
early years of production.

CRA would also wish to minimise uncertainty as
to future operating conditions, both to protect its
own interests and to allay its creditors' fears of
political instability. It would therefore attempt to
have its rights and obligations clearly specified,
and to obtain assurances that the agreement would
not be altered to the Company's detriment.

The Bougainville Copper Agreement, 1967

The terms of the arrangement between the Administration and Bougainville Copper Pty. Limited (BCPL) (a
CRA subsidiary incorporated in Papua New Guinea),
were contained in the Mining (Bougainville Copper
Agreement) Ordinance 1967.[33] BCPL would be granted a
Special Mining Lease for an initial period of 42
years, the principal financial terms of which would
be as follows. The Company (BCPL) would pay royalties at the rate of 1.25  per cent of the f.o.b.
sales value of minerals produced (Clause 5(h)). BCPL
would be exempt from income tax during the first

three years of commercial production (Clause 7(a)). Under general taxation provisions, the Company could allow for depreciation and amortisation and for financing costs in calculating its taxable income. Clause 7(a) and (b) provided that these allowances would not be claimed during the three year tax exemption, but would be accumulated and set against revenue in calculating taxable income from the fourth year onwards. This provision was necessary if income earned during the initial three years was to be permanently exempted from taxation; the accumulated allowances would be deductible immediately, and thus BCPL's liability to taxation would be postponed for some years after the end of the tax free period.[34] Clause 7(e) exempted 20 per cent of BCPL's net income from taxation, a concession then generally available to mining companies in Australia and Papua New Guinea and which the Company would continue to enjoy even if the general exemption were revoked.

When BCPL's net income did become liable to taxation, the applicable rate would rise from 25 per cent[35] to 50 per cent over four years (Clause 7(k)). After the twenty-fifth year of production the tax rate would rise by one per cent each year to a maximum of 66 per cent (Clause 7(p)). BCPL also undertook to offer the Administration 20 per cent of its equity at par value.

The **Agreement** also set out the rights and obligations of the two parties. The Company agreed to carry out a detailed feasibility study of the Panguna prospect as quickly as possible and, if the outcome was favourable, to construct within five years facilities for mining and transporting ore in commercial quantities (Clause 6(a)). In meeting its construction obligations, the Company would use supplies, machinery and equipment manufactured or produced in the Territory so far as was 'reasonably and economically practicable' (Clause 9(a)). The question of further processing of mineral output was left at the Company's discretion (Clause 16). BCPL also undertook that, so far as was 'reasonably and economically practicable', it would

> use and train in new skills labour available in the Territory and...continue and expand its training program...with a view to the early employment by it in technical and staff positions of suitably qualified inhabitants of the Territory (Clause 10(a)).

The rights of the Company were also spelt out in detail. It would of course receive Leases for mining and ancillary purposes (Clause 5(a)). It would have the right to import any items it needed (Clause 9(b)), and would be free to recruit abroad personnel required for construction or mining activity (Clause 10(b)). It could construct a port and roads on Bougainville Island; public access to these facilities would be permitted, except where BCPL felt that such access would interfere with safety or with mine efficiency (Clauses 11, 14). The Company was also granted the right to construct a hydro-electric power station and to generate, transmit and sell electric power, and it could take and use water for this and other purposes.

The Administration agreed to provide and maintain 'education, police, postal, telecommunication and medical facilities'. In general terms, its other obligations corresponded to the rights of the Company, while its principal rights were to levy the taxes and other charges outlined above and to obtain a 20 per cent equity shareholding in the operating company.

Another purpose of the Agreement was to guarantee BCPL against financial or other impositions beyond those specifically laid down in the Ordinance. It would not be disadvantaged by any alterations to the Territory's general tax legislation and would not be subject to any taxes or tariffs which were discriminatory in law or in practice (Clause 7); one effect of this provision was that import duties and a range of other taxes were frozen at the low levels which then applied. The Company would be free to choose its employees at all levels, its contractors, suppliers and customers, and to declare, credit and pay dividends. The Administration undertook not to 'expropriate or permit the resumption of any asset...of the Company'; the term 'expropriation' did not simply imply confiscation of property but would also include 'any substantial interference with the rights of the owner to control or carry on...[his] business...' (Clause 17).

Both the Administration and CRA wished to obtain House of Assembly ratification for the Agreement and the relevant Ordinance was discussed in the House on 29 and 30 August 1967. Little detailed comment was made on its terms, and very few Papua New Guinean members spoke. This was hardly surprising; many of the provisions were extremely complex, as was the language in which they were couched. Those members who did speak, including Paul Lapun, expressed general approval of the Agreement.[36]

## Assessing the Agreement: the Administration's Negotiating Effort

It is difficult to develop a standard against which to assess the Agreement. A number of authors have attempted to use similar agreements negotiated elsewhere at about the same time as a standard,[37] but this approach raises two major problems. First, it does not allow conclusions regarding the Agreement as a whole, since each arrangement contains a different mix of provisions relating to taxation, depreciation, government shareholdings and provision of infrastructure. For example, the arrangement concluded between Indonesia and Freeport Minerals Limited in 1967 for the development of the Ertsburg copper deposit provided for a lower nominal tax rate than did the Agreement, but did not permit immediate write-off of capital expenditure. The Indonesian government was not offered a shareholding in the operating company, but on the other hand did not undertake to provide and maintain a part of mine infrastructure.[38] The full financial ramifications of the two agreements could not therefore be compared until mining had been carried on under each for a substantial period of time.

Even if such a comparison was available it would not offer a basis on which to assess the Australian Administration's negotiating effort, for two reasons. First, the potential bargaining power of other host countries might have been different. Second, other host country governments might have failed to exercise the potential bargaining power which they did possess. A comparison between the Agreement and the minimum concessions required to attract CRA's investment would allow a judgement as to whether the Administration fully utilised its bargaining power, but it is difficult to establish what those minimum concessions were. CRA has not published any relevant information, while the absence of large-scale mineral developments prior to 1967 deprives us of what might have been a useful yardstick.[39]

The report of an international mission which studied development strategies for Papua New Guinea in 1972 (the Faber Report) claimed that the Agreement offered the original shareholders (CRA and related companies) a return far in excess of that required to attract the necessary investment. In support of this viewpoint it quoted a feasibility study carried out in 1969 which estimated that these shareholders would receive an annual return on capital of 34 per cent (at conservative metal price forecasts.) The Faber Report concluded that the Administration's over-generosity resulted in part from its lack of relevant expertise.[40]

The Report's emphasis on the feasibility study's profit forecast masks a very important point - the study was carried out in 1969. It could not have been compiled in 1967, because the basic operational characteristics of the prospective mine were unknown.[41] Consequently, the Australian negotiators could not gauge its likely profitability even in very general terms; the only guide they had was CRA's claim that the project would be marginal. Its lack of information placed the Administration at a serious disadvantage, and it could have done one of two things to strengthen its position. It could have negotiated a flexible agreement which would take account of the possibility that its bargaining position would improve as additional information became available, or it could have taken steps to immediately increase the level of information it possessed. It did neither, negotiating financial terms which were largely inflexible, and failing to obtain an independent assessment of the likely character of any development at Panguna. Thus its failure was not in neglecting to utilise its available bargaining power, but rather in neglecting to provide for, or to bring about, increases in that power.

It is important to note that the Administration could have negotiated a considerably more flexible arrangement while giving BCPL many of the assurances it sought regarding operating conditions. For example, renegotiaton of the Agreement could have been provided for if (and only if) the project proceeded on a scale considerably in excess of that anticipated, or a sliding scale of royalties could have been negotiated, the upper range of which would only apply if the project was unexpectedly profitable. And while it lacked the technical competence to assess the prospects for Panguna itself, the Administration could have employed independent consultants to do so. But it chose to accept CRA's projections at their face value, and thus displayed considerable naivety since it was extremely unlikely that the Company would describe its deposit as anything but marginal.[42]

Was the Administration engaged in complicity with CRA? The explanation for its failure to negotiate more flexible financial terms apparently lies more in its low position on the relevant 'learning curve'. The Administration did not, in fact, lack expert advice per se, as it was assisted by staff employed by Australia's Federal Government in its Treasury Department, its Attorney General's Office and its Bureau of Mineral Resources. However, the Administration's claim that these officials 'were experienced in making similar agreements in Australia'[43] was probably incorrect. The Australian Federal Government did not then negotiate terms for the development of particular projects, but rather applied a general tax regime to all mining operations.[44] Their professional background may have led the Administration's advisers to ignore the importance of, and the need to provide for, shifts in the relative bargaining powers of the host country and the foreign investor in relation to specific projects.

The Administration's acceptance of CRA's profit forecasts is a different matter. Its uncritical approach may have reflected a failure on its part to distance itself sufficiently from the fellow Australians who ran CRA. It is evident from the statements of its spokesmen in the House of Assembly and from the policies it implemented on Bougainville that the Administration perceived a considerable degree of affinity between its interests and those of CRA, a perception which can only have been enhanced when it obtained a 20 per cent holding in the Company. This feeling of affinity may have dulled the Administration's awareness that its interests differed from CRA's in important ways, and that the Company might have a strong incentive for being less than frank with it. On the other hand, the Administration may have simply been overawed by and unwilling to challenge CRA's obvious superiority in the field of mining expertise, a feeling it would certainly have shared with authorities in many other host countries.

Preparations for Mine Development

Having concluded an agreement with the Administration, BCPL carried out a detailed feasibility study of the Panguna prospect. It was concluded that an open pit operation on as large a scale as possible was required and that mining should initially be concentrated on the high-grade sections of the

orebody.[45] Physical, financial and marketing
constraints indicated that approximately 90,000 tons
of ore per day should be mined, inferring annual
metal production (in concentrates) of about 165,000
tons copper, 500,000 ounces gold and 1 million
ounces silver. Production would therefore be far in
excess of that anticipated by the Administration
when it negotiated the Agreement (70,000 tons
copper, 200,000-300,000 ounces gold).[46]

Market outlets and loan finance would have to
be secured before development could proceed. It had
been decided to dispose of Panguna's output in
concentrate form, and BCPL successfully negotiated
long-term contracts with smelters in Japan, Germany
and Spain providing for sales of concentrates with a
total copper content of 1,992,500 tons over fifteen
years. BCPL would be paid for the metal content of
its concentrates at prices based on the relevant
London metal exchanges. BCPL required an estimated
US$250 million in loan capital to help finance the
planned development at Panguna, and in early 1969
CRA and RTZ set about raising this amount. A consortium of 27 banks from 8 countries was assembled, and
in July 1969 a Credit Agreement was signed between
it and BCPL. Two separate loans were negotiated. The
first, known as the Intermediate Loan, was for a
maximum of US$154 million, a sum which would be made
available directly to BCPL. The second, or Long Term
Loan of US$92 million would be provided to the
Commonwealth Trading Bank of Australia (CTBA), which
would then pass on an equivalent amount in
Australian dollars to BCPL.

The loans were guaranteed by a pledge of BCPL's
shares, which effectively meant that CRA was the
principal guarantor. RTZ was not a guarantor, but it
undertook to maintain a controlling interest in BCPL
at least until repayment was complete.[47] RTZ played
a crucial role in raising finance for BCPL. It
utilised its extensive links in the international
banking community to assemble the Consortium, while
the Consortium's willingness to provide finance
depended to an important degree on RTZ's assurance
that its technical and financial expertise would be
available to CRA and BCPL.

The necessary financial and marketing arrangements were completed by July 1969, and construction
was soon under way on Bougainville.

Renewal of Conflict and the Emergence of
Secessionism on Bougainville

As CRA's activities expanded and its presence took
on an air of permanence, some Bougainvillians began
to consider more radical political action, particularly the possibility of secession. Secessionist
sentiment existed on Bougainville before the
Company's arrival,[48] but it was only after 1964 that
such sentiment was articulated by politically-
influential individuals, or was widely reflected in
public opinion. Many local people believed that the
Administration was deliberately sacrificing their
interests in pursuit of its own and CRA's objectives, and that Bougainvillians would bear the costs
of mineral development while other Papua New
Guineans enjoyed the benefits. To people who felt
threatened and abandoned, secession seemed an
attractive alternative, and the prospect of substantial copper revenues appeared to make it a
financially viable one.
   Secessionist sentiment was strengthened by
renewed conflict over land in 1968-9. By mid-1968
the Administration and CRA were looking for land
near Kieta on which to construct a town and other
facilities. A number of clashes occurred between
local people and Company surveyors, one of which led
to the arrest and imprisonment of villagers from
Pakia. The Administration examined a number of
potential town sites, and chose an area at Arawa
comprising a 1,000-acre expatriate-owned plantation
and 650 acres owned by the Arawa people. One hundred
and forty five acres of the Rorovana people's land
was earmarked as a location for a construction camp
and other facilities; this land would be leased by
CRA, while that at Arawa would be alienated permanently.
   The Administration initiated negotiations for
purchase of the land; the plantation's owners
accepted its terms after a long struggle (in which
they were supported by the local people) and under
threat of compulsory acquisition. The Arawa people
rejected them, as did some Rorovanas, and the
Administration obtained a compulsory leasing order
on the Rorovana land. The villagers ignored the
order, and in August 1969 riot police employed
physical force (including use of tear gas) to remove
them. The Administration's action received wide
publicity, much of it critical, in Australia. The
Australian Prime Minister, Mr Gorton, intervened in
the dispute, and instructed the Administration to

depart from its principle of not allowing direct negotiations between local people and CRA. This principle reflected the Administration's determination that land sales on Bougainville should be at prices applicable in the Territory as a whole, so as to avoid setting a precedent which might be applied to purchases of land needed for development elsewhere. It believed, correctly, that CRA would pay prices well in excess of the norm in order to avoid costly delays.

CRA soon negotiated a 42 year lease with the Rorovanas on terms considerably more favourable than the Administration had offered, and a similar arrangement was subsequently concluded between the Administration, BCPL and the Arawa people.[49] Having been granted a much more generous settlement, the coastal people accepted the Company's presence and there was no further violence. But the threats of compulsory acquisition and the use of force at Rorovana lost the Administration whatever credibility it retained with local people.

The Administration had attempted to implement two policies which were in practice irreconcilable. It tried to ensure a rapid pace of development, while refusing to grant villagers compensation above the agricultural value of their land. But if development was to proceed rapidly, concessions had to be made on the issue of compensation. The alternative was conflict.

In June 1969 the Director of District Administration, Mr Ellis, outlined to the House of Assembly actions which the Administration had taken over the previous two years to improve communication with Bougainvillians. He detailed work undertaken by officials in informing local people of CRA's plans, in explaining the Company's rights and the villagers' entitlement to compensation and other payments, and in accompanying prospecting teams into villages. It is apparent from the information supplied by Ellis that the Administration invested considerable time and effort in this work.[50] But it clearly failed to understand that one action such as the use of riot police at Rorovana could counteract the years of painstaking effort.

By early 1972, preparation of the Panguna deposit was complete, processing facilities were installed, and sites prepared for the disposal of waste rock and mine tailings. Town sites complete with recreational, commercial and educational facilities had been constructed at Panguna and Arawa, a two-lane highway connected Panguna to the coast, and port, power generation and fuel storage facilities were established at Loloho (near Arawa). The Administration had upgraded the Aropa air strip, constructed a road from Kieta to Arawa and Loloho, installed a telecommunication system and constructed schools, houses and other facilities at Arawa. Total project costs plus working capital exceeded US$480 million. Of this sum, loan capital accounted for US$334 or A$282 million, while BCPL's shareholders provided a total of A$133 million or US$151 million. BCPL commenced commercial production on 1 April 1972.

Notes

1. Far Eastern Economic Review, Asia 1979 Yearbook (Hong Kong, 1979), p. 277.
2. J. Momis and E. Ogan, 'A View from Bougainville', in M.W.Ward (ed.), Change and Development in Rural Melanesia (Australian National University Press, Canberra, 1972), pp.106-18.
3. Ibid., p. 108.
4. What information does exist is summarised by M.L. Treadgold, The Regional Economy of Bougainville: Growth and Structural Change, Development Studies Centre Occasional Paper No. 10 (Australian National University, Canberra, 1978), pp. 2-16.
5. In the late 1960s, Bougainville accounted for one sixth of PNG's copra production and one quarter of its cocoa production.
6. TPNG, Bougainville Draft Development Programme (Port Moresby, September 1969), p.4.
7. Use is made of Phillips' own account of events in this section: see K. Phillips, 'Notes on Contact with Bougainvilleans in the Early Exploration Phase', reproduced in R. Bedford and A. Mamak, Compensating for Development: The Bougainville Case, Bougainville Special Publications No.2 (Christchurch, 1977), pp. 151-60. Phillips' account was verified where possible in interviews held with Severius (Sev) Ampoai, a local schoolteacher who was closely involved in events during this period, at Panguna on 19 June 1978.
8. Phillips, in Bedford and Mamak, Compensating for Development, p.156.
9. Ibid., p.157.
10. TPNG House of Assembly Debates, 8 June 1966.
11. Ibid.
12. Ibid.
13. Ibid., 14 June 1966.
14. Ibid.

15. Ibid., 24 June 1966.
16. This has been confirmed by Sev Ampaoi, who during these years was closely involved in community affairs (as a schoolteacher and as chairman of the Local Government Council), and who would certainly have known of any such attempt. Interview, Panguna, 19 June 1978.
17. Quoted in J. Dove, T. Miriung, M. Togolo, 'Mining Bitterness', in P.G. Sack (ed.), Problems of Choice: Land in Papua New Guinea's Future (Australian National University Press, 1974), p.182.
18. See, for example, TPNG House of Assembly Debates, 14 June, 23, 24 November 1966.
19. Quoted in Bedford and Mamak, Compensating for Development, p.21.
20. See, for example, the statements by Grove and by Newman, TPNG House of Assembly Debates, 23 and 24 November 1966.
21. Phillips in Mamak and Bedford, Compensating for Development, p.157.
22. Ibid., p. 158.
23. 'Report of the 1962 United Nations Visiting Mission to the Trust Territory of New Guinea' (New York, 1962), pp.24-6.
24. International Bank for Reconstruction and Development (IBRD), The Economic Development of the Territory of Papua New Guinea (Johns Hopkins Press, Baltimore, 1965), pp.31-2.
25. Interview with Sev Ampoai, Panguna, 19 June 1978.
26. TPNG House of Assembly Debates, 14 June, 24 November 1966.
27. Ibid., 14 June 1966, 23, 24 November 1966.
28. In fact a proportion of royalties could have been paid to villagers without recognising their claims to mineral ownership, for example by calculating occupation fees as a percentage of the value of mineral production.
29. All applications for licences were considered before a public Warden's Court, but in fact not one of CRA's applications was rejected.
30. This feeling was further enhanced when the Administration purchased a 20 per cent shareholding in CRA's Panguna operation.
31. The Administration failed to realise that its decision to forego mineral revenues would itself reduce the Panguna mine's impact on the balance of payments.
32. The most important of these related to the low grade of the Panguna ore, to the scale of operations which would be required to exploit that ore profit-

ably, and to the price instability which characterises world copper markets.

33. TPNG, Ordinance No. 70 of 1967 (Government Printer, Port Moresby, January 1968).

34. The exact duration of the postponement would depend on the scale of relevant capital expenditures and on the Company's profitability.

35. This was the standard rate in the Territory at the time.

36. TPNG House of Assembly Debates, 29, 30 August 1967.

37. See for example, G.O. Guttman, 'Objectives, Strategy and Tactics in Negotiations for Mining Enterprises', in J.G. Zorn and P. Bayne (eds.), Foreign Investment, International Law and National Development: Papers presented at the Seventh Waigani Seminar (Sydney, 1975), pp.102-5; McKern, Multinational Enterprise, pp.90-1.

38. For details of the Ertsburg Agreement, see Guttman in Zorn and Bayne, Foreign Investment, p.104; McKern, Multinational Enterprise, p.90.

39. In the Irish case, for example, it was possible to point to mineral developments which had previously taken place under less generous provisions in claiming that the granting of a twenty year tax exemption was unnecessary.

40. Overseas Development Group, University of East Anglia, A Report on Development Strategies for Papua New Guinea (Port Moresby, 1973), p.69.

41. D. Mentz, 'Administrative Aspects of the Bougainville Development', in Papua New Guinea Society of Victoria, Background to Bougainville - a Factual Analysis (Melbourne, 1969), p.12.

42. As a United Nations official with wide experience of assisting host governments in negotiating mining agreements has stated, '...I have never been in a negotiation where the project has been described by the mining company other than as a marginal deposit'. C.J. Lipton, 'Government Negotiating Techniques and Strategies', in United Nations, Negotiation and Drafting of Mining Development Agreements (Mining Journal Books, London, 1976), p.107.

43. TPNG House of Assembly Debates, 30 August 1967.

44. State governments negotiated items specific to individual developments such as royalty rates and terms for provision of infrastructure.

45. The Panguna deposit was characterised by significant variations in ore grades. While the cut-off grade was set at 0.2 per cent copper, two ore zones grading in excess of 0.6 per cent copper were

identified.
46. TPNG House of Assembly Debates, 8 June 1966.
47. BCPL, 'Credit Agreement, 28 July 1969', unpublished, Appendix f.
48. J. Ryan, The Hot Land: Focus on New Guinea (Macmillan, Melbourne, 1969), pp.335-6.
49. Bedford and Mamak, Compensating for Development, p.26.
50. TPNG House of Assembly Debates, 19 June 1969.

Chapter Nine

MINERAL DEVELOPMENT IN POST-COLONIAL PAPUA NEW GUINEA, 1972-1980

## Political Developments 1969-72

By April 1972 important political changes had occurred in Papua New Guinea. Throughout the 1960s it was generally assumed in Port Moresby and Canberra that Australia would not take the init- iative in granting self-government or independence until requested to do so by Papua New Guinea's political leaders. However during these years indigenous politicians displayed little impatience to take up the reins of government. In 1969 the leader of the Australian federal opposition, Mr Whitlam, made a number of statements which repres- ented an important departure from the previously bipartisan policy. He stated that his Labor Party would grant immediate self-government if it won power in the 1972 elections, and would offer indep- endence by 1976. Whitlam's commitment exerted considerable influence on Australian government policy, and in April 1971 the Minister for External Territories announced that a 'leadership group' would be chosen by the next House of Assembly. This group would exercise powers already delegated to Papua New Guinea and would negotiate self-government with Canberra.

The new House of Assembly was elected in March 1972. It had been expected that the conservative United Party would win a majority in the House, but in fact the more radical Pangu Pati joined with a number of smaller groupings to form a coalition government in April 1972. Its Chief Minister was Mr Michael Somare and its Minister for Mines, Mr. Paul Lapun.

Among the newly-elected members was a Bougain- villan priest, Fr John Momis, who was strongly critical of BCPL's activities and of the **Agreement**.

On 27 September 1972 Momis introduced a motion which
proposed that government policy towards foreign
mining investment be based on certain principles.
These dealt, among other things, with government
equity participation in mining ventures, employment
and training, generation of linkages, and taxation
of mining company profits. But the principle of
greatest immediate significance was that mining
agreements 'should make provision for [their]
renegotiation in the light of changing circumstances
and the acquisition of knowledge'.[1] The government's
official position on the possibility of renegotiation remained unclear. While Mr Somare stated that
'we are not questioning previous agreements with the
Administration', Paul Lapun's view was that 'if at a
later date the Government sees that the company is
making colossal profits I think the Government
should be able to renegotiate ...'[2] The United Party
opposition criticised suggestions of a renegotiation, arguing that the Agreement should be
honoured regardless of whether it was a good one or
not. An 'officially and legally signed contract'
existed, and breach of that contract would destroy
Papua New Guinea's 'good name' and its standing with
the international financial community, deterring
foreign investment in mining and other industries.[3]
Despite these reservations the opposition gave its
general support to the motion, which the House
passed on 23 November 1972.

Development Strategies and the Bougainville Copper
Agreement

The development strategy pursued by the Administration in the years prior to 1972 was based on the
1963 World Bank Report, which recommended monetary
sector expansion based largely on production of
agricultural and forestry goods for export. The Bank
advocated that 'expenditures and manpower be concentrated in areas and on activities where the prospective return is highest'. Resources should be chann-
elled into 'areas of good land which are readily
accessible and where development is relatively
easy'; once the economy had grown, allocation of
resources to more difficult areas could be
considered. It urged the Administration to 'concentrate its efforts on the advancement of the native
people',[4] but did not question whether its economic
policies were appropriate in the light of this
recommendation.
    In 1968 the Administration outlined its

strategy in a document entitled _Programmes and Policies for the Economic Development of Papua New Guinea_.[5] This document called for 'considerable expansion in the agricultural, pastoral, forestry, mining and manufacturing industries, requiring heavy private investment and expatriate involvement'. Private investment would come substantially from expatriate businessmen and foreign companies who alone would possess the necessary financial and technical resources. The aim of this development strategy was to achieve 'the maximum participation of the indigenous people as a means of progress towards economic self-reliance and social advancement'.[6]

In this instance, it was very apparent that discrepancies might exist between stated policy aims and the measures advocated for their achievement. Since expatriates and foreigners would provide the main impetus for growth, they would presumably also reap the direct benefits of economic expansion. This raised the question of precisely what kind of participation indigenous people would enjoy and to what extent real social advancement would occur. The Administration's position was based on two important assumptions, first that if aggregate income grew quickly the benefits would automatically 'trickle down' to all Papua New Guineans, and second that indigenous people would be better off 'with a small share of a large cake because at Independence or some time thereafter it is the size of the cake that counts, not who owns what share'.[7]

The political developments described above were accompanied by changes in the climate of opinion as to what constituted appropriate development strategies for Papua New Guinea. An important milestone in this regard was the 1972 _Faber Report_. This rejected the basic assumption of the Administration's economic policies, i.e. that rapid monetary sector growth would automatically enhance the economic and social well-being of Papua New Guineans.

The evidence available in 1972 indicated that the economic situation of nationals was worsening in relative terms. While monetary sector product grew at an annual rate of 15.6 per cent over the period 1966-70, the share of monetary sector income accruing to the indigenous workforce fell from 36.0 per cent to 32.0 per cent.[8] Increased reliance on foreign investment was partly responsible for this trend; the proportion of national income accruing to foreign companies increased from 5 per cent in 1967/8 to 16 per cent in 1971/2. Dependence on expatriate labour also played a part; the proportion of household income accruing to nationals fell from 79.5 per cent in the early 1960s to 63.4 per cent in 1970.[9] Papua New Guineans were, of course, receiving a smaller slice of a substantially larger cake, but the Faber Report argued that any relative decline in indigenous incomes was likely to cause political and economic problems.[10] In other words, it mattered very much 'who owned what share of the cake'.

Administration policy had other weaknesses. Concentration of resources in the most productive sectors of the economy offered no solution to, and probably accentuated, inequalities in regional income distribution. The rapid growth of the monetary economy was not matched in the subsistence sector; a rapidly-growing imbalance was emerging between the two sectors, an imbalance associated with substantial migration from rural to urban areas. Employment opportunities in the towns did not match the inflow of labour.[11]

Clearly, economic growth per se did not guarantee a solution to Papua New Guinea's problems and might well add to them. The Faber Report set out to identify social, political, and economic priorities and to seek the type of investment and growth patterns most likely to facilitate their implementation. The priorities it identified were to increase indigenous control of the economy, expand opportunities for income-generating self-employment for indigenous people, increase income to nationals while avoiding large disparities in local income distribution, reduce dependence on foreign aid and enhance prospects for rural development.[12] Regional development and rural revitalisation were seen as indispensable if other policy aims were to be achieved. The urban cash economy could not grow quickly enough to supply sufficient employment for Papua New Guinea's expanding workforce, and so employment opportunities in rural areas would have to be enhanced. Rural development could be expected to increase indigenous incomes while reducing dependence on foreign aid and investment in the longer term, thereby increasing indigenous control of the economy.

The principal role of enclave projects would be to provide revenue which government would use to help poorer regions catch up on lost ground and generally to facilitate and encourage indigenous economic activity. Given this general approach, it was hardly surprising that the Faber Report should state that 'earlier and substantially greater contributions to public revenue should be obtained from the Bougainville project'.[13]

Policies of the type advocated by the Faber Report were very much more attractive to the Papua New Guineans who were taking over power than a strategy which seemed to promise many of the rewards of economic development to foreigners while failing to deal with increasingly apparent economic and social problems. The policy priorities outlined by Mr Somare's Cabinet in December 1972 were based closely on the Faber Report's recommendations.

By 1972 two of the major assumptions which underlay the Australian Administration's negotiation of the Agreement were no longer valid. First, continued Australian subsidisation could not be guaranteed now that independence was at hand, and in any case the national government was anxious to avoid dependence on Australian aid. Second, it was felt that enclave projects would not directly create significant benefits for Papua New Guineans, and should therefore generate substantial government revenues at an early date.

The Background to Renegotiation

Throughout 1973 a number of politicians, particularly John Momis and Paul Lapun, continued to press for a critical reappraisal of the Agreement. Meanwhile support for such a move was growing among the staff which the Somare government had recruited to replace Australian Administration officials, though a wide spectrum of opinion existed as to how, or to what extent, the Agreement should be altered. The two main strands of opinion consisted of those who wished to pressure Bougainville Copper Limited (BCL)[14] into providing more revenue immediately without substantially changing the Agreement, and those who wanted a full-scale renegotiation of its terms.

In May 1973 the Finance Minister, Julius Chan, requested one of his staff, Ross Garnaut, to suggest a tax system that would 'take as much money as possible from (BCL) consistent with similar arrangements being applied to new mineral investments'.[15] Garnaut prepared a proposal in conjunction with a University of Papua New Guinea (UPNG) economist, Anthony Clunies Ross.[16] Their basic assumption was that host countries should obtain as high a proportion as possible of mining company profits after deduction of the company income which would correspond to the minimum return necessary to attract investment into new projects. In designing a tax system for this purpose, two considerations would be particularly important. First, most host countries would not possess the information on production costs and product prices needed to permit an accurate forecast of project profitability. Thus a single appropriate tax rate could not be fixed in advance. Second, host country authorities would wish to ensure that investment in marginal deposits was not deterred while making certain that profitable projects would generate substantial government revenue.

Garnaut and Clunies Ross recommended use of a Resource Rent Tax (RRT), that is 'a profits tax that begins to be collected when a certain threshold rate of return on funds invested ... has been realised'.[17] RRT would be collected at progressively higher rates as higher rates of return were realised. Under such a system, a company contemplating investment in what appeared to be a marginal project would know that little or no tax would be incurred until it had earned an 'appropriate' return. If the project was unexpectedly profitable, RRT would apply, ensuring that the government captured a large proportion of windfall profits. Such a system would minimise the cost to government of its lack of information, since a single tax rate would not have to be set in advance. Neither would absence of such a tax create uncertainty for investors, since they could calculate precisely what tax levels would apply given various assumptions concerning costs and prices.

Application of RRT would of course necessitate a substantial renegotiation of the Agreement. An alternative approach was advocated by Louis T. Wells, of the Harvard Business School, whom the government invited to examine the Agreement and recommend any changes he considered desirable. Wells advised a cautious approach. In his view, mining agreements in Papua New Guinea's immediate geographical area were less stringent than in other parts of the world, and so it would have to be careful not to price itself out of the market. Wells suggested that additional revenue could be obtained immediately by abolishing the tax exemption on 20 per cent of BCL's income and by beginning write-offs of capital expenditure during the first year of the tax holiday rather than at its completion. He opposed imposition of a profit-based tax in addition to standard corporation tax.[18]

Throughout late 1973 political momentum in

favour of renegotiation gathered force while the debate among proponents of the various strategies continued. BCL itself helped to decide the issue in early 1974 by announcing net profits for 1973 of A$158.4 million. The government would receive about A$29 million of this amount, mostly as dividends on the equity it had purchased, and from dividend withholding tax (introduced in the 1972 budget at a rate of 15 per cent) on dividends paid to non-nationals. It found this situation completely unacceptable, and decided to proceed with a full-scale renegotiation of the Bougainville Copper Agreement.

Preparing for the Renegotiation

The government devoted considerable effort to the task of preparing for negotiations. It already employed competent advisers on general economic and legal matters, and it now recruited specialised staff (usually on a short-term basis) in those areas where it lacked expertise. Assistance was received from the Commonwealth Secretariat and the Intergovernmental Council of Copper Exporting Countries (CIPEC). Few of the individuals involved were to remain in Port Moresby throughout the negotiating process, but their presence was of considerable help to the government in formulating its negotiating position.

Another important step was the establishment and maintenance of liaison between senior politicians and advisory staff. Mr Somare had earlier formed a ministerial committee to consider the whole question of renegotiation; staff members now reported to this committee regularly, and care was taken to keep it fully informed of all relevant developments. In addition, the renegotiation was periodically discussed by Cabinet, and the basic principles being applied were reaffirmed. The government was anxious that Bougainvillians should feel that they had been represented at the renegotiation, and that its position should not be weakened by conflict with the islanders. It therefore included John Momis and Leo Hannett (a prominent political figure on Bougainville) in the team which would handle the renegotiation.

But perhaps the most important preparatory work was carried on as part of the general process of policy formulation undertaken between 1972 and 1974. The politicians and civil servants involved rejected many of the Australian Administration's assumptions,

and in doing so articulated the fundamental principles which would underlie their own approach. As part
of this process, agreed priorities were established
concerning the role which mining projects would be
expected to play. Consequently the government knew
exactly what it wanted (and was united in its commitment to obtain it) when negotiations commenced.

The broad policy aims of Mr Somare's government
have been mentioned above. They dictated that 'the
government should be in the business of rural
development, not of mining', as one official
expressed it.[19] Consequently, its priority would be
to maximise mineral revenues which could then be
used to finance rural development; it would not
demand increased ownership of BCL nor, with some
exceptions, would it seek increased control over the
Company's operations. The government did not wish to
increase its shareholding in BCL for a number of
reasons. First, it believed that its limited administrative capacity should be used to foster rural
development rather than to oversee a mining concern.
Second, a substantial commitment of funds would be
required,[20] and it believed that investment in an
already-operating mine would represent an inefficient use of Papua New Guinea's limited financial
resources. In addition, it was felt that CRA and RTZ
would have little incentive to commit their technical and financial resources to the Panguna project
if the state substantially increased its shareholding. Increased state ownership was not seen as a
prerequisite for capturing a higher share of BCL's
profits; an efficient tax system would do this
without involving the state in the expense (and the
risk) of further equity investment.

The government believed that Papua New Guinea
would not be disadvantaged if BCL retained complete
control of many aspects of mining operations. For
example, there was little concern with the Company's
marketing practices, since it did not sell concentrates to affiliates and could therefore be expected
to market its products at the highest available
price. Neither was there any desire to influence
corporate decisions affecting the labour intensity
of mining operations. The government was not concerned with creating additional employment on Bougainville, since this could only lead to further immigration and increased social and political tensions.
Nor did it wish to influence BCL's purchasing
policies. It felt that the domestic economy was not
in a position to supply the goods and services
required, and that it was not part of its develop-

ment strategy to help establish industries which
would probably be uncompetitive in any case.[21]
Underlying this relative disinterest in the direct
economic impacts of mining (employment, linkages)
was the belief that their effects would be geographically concentrated and that the type of development they would represent might frequently be
undesirable. Mineral revenues, on the other hand,
could be expended for the benefit of all Papua New
Guineans and so as to foster desired types of
development.

In addition to the question of revenue generation, two other issues would receive attention. The
government would wish to ensure that BCL's training
and localisation policies would maximise indigenous
participation in the Company's activities, and that
the adverse environmental effects of the Panguna
mine on surrounding rural areas be minimised. Thus
when negotiations commenced the government had identified a well-defined and limited range of issues on
which its negotiators would concentrate their
attention.

The Conduct of Negotiations

Negotiations took place over the period May to
September 1974. BCL rejected the government's first
proposal out of hand. The basic elements of this
proposal were to be contained in the final settlement, and BCL's attitude at this stage apparently
reflected its desire to test the government's determination.[22] In June a second proposal was submitted
to the Company, and it met this with a counterprop-
osal which was discussed at a short meeting in Port
Moresby. BCL's terms were rejected by government
negotiators, who felt that the changes proposed were
cosmetic in nature and were intended to present the
Company as willing to compromise while leaving the
division of profits essentially unchanged. The
government was angered at what it saw as BCL's
refusal to enter into serious negotiations, and it
informed the Company that the Agreement would be
amended by legislation if a satisfactory arrangement
was not negotiated. Just before the Port Moresby
meeting broke up, BCL submitted another proposal for
consideration, and its terms made it clear that the
Company was now ready to negotiate.[23]

In August 1974 the main phase of negotiations
began, and agreement was soon reached on a number of
non-financial matters. The government was anxious to
compile an environmental impact study of the Panguna

project and BCL agreed to assist in this work.[24] The Company undertook to commission a pilot study which would examine the economic feasibility of smelting or refining copper concentrates in Papua New Guinea.[25] BCL's localisation and training policies were also discussed, and the government accepted that progress was generally satisfactory. The parties agreed to keep in close consultation on the matter and the government concluded that this, combined with its powers under generally-applicable legislation, would allow it to exercise its influence whenever necessary.[26]

Argeement was not so easily achieved on financial matters. The government's demands were as follows: abolition of the tax exemption on 20 per cent of BCL's income; termination of the three-year tax exemption from 1 January 1974; withdrawal of accelerated depreciation allowances; and imposition on BCL's profits of corporation tax at the standard rate (then 33 per cent) and of an Additional Profits Tax (APT) which would yield a total tax rate of 70 per cent on that portion of profits above a sum which would represent a return to BCL of 15 per cent on funds invested.

Financial terms were finally agreed upon in September after two protracted negotiating sessions. Previously much of the actual bargaining had been between advisory staff and executives of CRA and BCL, but by this stage the most senior personnel on both sides were involved. The Company team was led by Sir Val Duncan, the RTZ Chairman, and CRA's Chairman, Rod Carnegie. The Papua New Guinea team included Paul Lapun, Julius Chan (Minister for Finance), and Maori Kiki (Minister for Foreign Affairs), and was headed by Mr Somare during the second session. At the first meeting Duncan suggested that the government's proposals might be acceptable if it agreed to buy a further 50 per cent of BCL's equity (at a substantial premium above the market price), a proposal which the Papua New Guineans rejected. They saw it as a diversionary device whereby RTZ could take back with one hand what it had just conceded with the other.

The government team proceeded to negotiate on the basis of its own proposals, applying 'a carefully orchestrated approach', as one witness expressed it, with each member playing a particular role. Mr Somare explained what he saw as the political realities and stated that his government's position would be undermined if it did not obtain a more favourable agreement. Mr Chan outlined Papua

New Guinea's financial situation and claimed that the country simply had to obtain additional revenue. Paul Lapun spoke in Pidgin of the costs which the mine had imposed on his people. Kiki took a very hard line throughout, demanding that all the government's requests be met and rejecting any compromise; thus he constantly reminded the Company of what it stood to lose if it refused to compromise.[27]

This 'orchestrated approach' was apparently not planned; rather the roles of individual ministers emerged from the interaction and discussion which occurred in numerous meetings of the ministerial committee and of Cabinet.[28] It was certainly effective. Gradually the Company conceded on the major issues, abandoning its 'non-negotiable principles' one by one.[29] It accepted the principle of APT, but Duncan fought long and hard to have BCL's 1974 earnings exempted from tax. Much was at stake here. BCL's first half operating profit amounted to $118 million, and under the proposed tax system the government would receive some $66 million of this. Finally, it was agreed that BCL's 1974 income would be liable to taxation, but an element of compromise was included in that first half profits would be taxed at a rate of only 44 per cent.

A number of additional points were also agreed upon. BCL could claim an increase in the return allowed on its funds before APT applied if 'abnormal conditions of inflation' occurred or if it was substantially disadvantaged by re-alignments in currency conversion rates. It undertook to pay 50 cents per tonne of copper sold into a Non-Renewable Resources Fund which would be expended for the benefit of Bougainvillians. Finally, the two parties agreed to meet in 1981 and every seven years thereafter to review the operation of the Agreement.[30]

The renegotiation in itself had little effect on the status of Bougainvillians in relation to BCL's operations, though they would receive some additional financial benefit from the Non-Renewable Resources Fund, while the involvement of Momis and Hannett in the negotiating team was a recognition of their special interest in the outcome. However, during the following two years the situation was to change quite dramatically. In December 1974 the national government agreed that all mineral royalties from Panguna should flow to Bougainville. In 1974-6 a provincial government was established on the Island and this very much increased Bougainvillian participation in decisions affecting the

mine. The provincial government is consulted both by
BCL and by Port Moresby regarding all major relevant
policy initiatives, and it will effectively have a
power of veto over any new mineral developments on
Bougainville.

Assessing the 1974 Agreement

The 1974 Agreement embodied nearly all the government's objectives; the only substantial
concession it made was to modify its demands
concerning BCL's tax liability for 1974, by about
A$7 million. What accounts for its success? The
manner in which the government prepared for negotiations was clearly very important, particularly in
that it identified limited and specific objectives.
This placed it in a strong bargaining position as it
could immediately reject any proposal which did not
accord with its aims, and its own proposals would
again become the basis for negotiations. This was
illustrated, for example, when the Company suggested
that Papua New Guinea buy a majority shareholding in
BCL. The government had already excluded this alternative, and could therefore swing negotiations back
to the question of tax revenue.
      The liaison which existed between advisory
staff and senior politicians was also of considerable importance. Over a period of time, these
politicians were 'educated' as to the issues
involved by their advisers, and a feeling of trust
and mutual confidence developed between the two
groups. When the important final negotiating
sessions occurred, the politicians were well
informed and working closely with the advisory
staff, and so could throw all their weight behind
the negotiating effort.
      The personal qualities of the individuals
concerned also contributed to a successful outcome.
The Advisory staff were technically competent in
their particular fields and also displayed a keen
tactical sense, a quality apparently scarce among
many groups of this kind. Their treatment of Sir Val
Duncan, the RTZ Chairman, is illustrative in this
regard. Duncan made a practice of dealing with the
countries in which RTZ operated at the prime ministerial level only, and on his arrival in Port
Moresby he asked to see Mr Somare, a request the
Prime Minister refused. Duncan persisted, but
officials and ministers ensured that there was
always some obstacle to a meeting, and the two men
did not come face to face until the final

negotiating session. This action apparently dealt a considerable blow to Duncan's confidence, and also informed him in no uncertain terms that the PNG government could not be treated lightly.

The quality of political leadership was also high. Unity was maintained throughout, and no ambiguity existed as to the commitment of any minister to the pursuit of Papua New Guinea's interests. The politicians most directly involved were, as a number of their advisers have attested,[31] open-minded, willing to listen to advice and rational argument, and ready to try alternatives which did not correspond to the conventional wisdom on relevant issues.

Many of the factors discussed above could not have applied had the general context of policy-making in Port Moresby not been characterised by a high degree of rationality. A wide range of alternatives was usually considered in each policy area, their implications were discussed and analysed, and choices were then made on the basis of Papua New Guinea's perceived interests. It was this climate which allowed the government team to make its preparations and which ensured that it enjoyed enthusiastic political support for its negotiating effort.

A number of factors underlay this climate of decision-making. The politicians involved felt that their actions would have a major impact on Papua New Guinea's history, and that a rigorous standard of decision-making should therefore be applied. Political development in Papua New Guinea was not far advanced, and consequently few organised pressure groups were attempting to influence decision-makers. In this context, a point of particular relevance was the absence of domestic groups with a vested interest in protecting BCL's profitability. Since the bureaucracy in Port Moresby was new, neither had vested bureaucratic interests emerged. Unlike many colonial regimes, Australia did not create a bitter legacy of racial antagonism,[32] and foreign companies could be objectively assessed in the light of their contribution to Papua New Guinea rather than condemned out of hand as agents of continued racial exploitation.

The attitude and actions of BCL/CRA/RTZ also played an important role in determining the outcome of events. BCL did not attempt to undermine the government's political position, or to mobilise external forces in support of its own position. If BCL had acted otherwise, it might of course have placed a A$500 million investment in jeopardy, and this may partly explain its cautious approach. Because of this approach, BCL would exercise whatever bargaining powers it possessed in the negotiating room and not outside it. Those powers were limited by a number of important constraints. First, the government possessed an ultimate sanction: if necessary, it could simply legislate to obtain the desired outcome.[33] Second, BCL could not credibly threaten to abandon a A$500 million investment while the government allowed it to receive a reasonable return on that investment. BCL could only argue for as generous an interpretation as possible of the term 'reasonable'. Third and more generally, circumstances in 1973-4 did not favour BCL's bargaining position. It was operating a successful mining venture during a period of high metal prices, and it was consequently difficult for the Company to defend a three-year tax exemption, especially given Papua New Guinea's transition to Independence and the accompanying changes in attitude towards foreign mining investment.

The taxation arrangements negotiated with BCL were subsequently used by the government as the basis for formulating a generally-applicable taxation regime for large mining projects, with, however, a number of significant changes being introduced in the process. In particular, APT is applied to all profits once the mining company has recovered its capital plus a designated rate of return, provision is made for application of accelerated depreciation allowances where the cash flow during an investment recovery period is less than a quarter of the initial capital expenditure, and carrying-forward of losses is provided for in assessment of APT.[34]

These arrangements appear to strike a reasonable balance between the need to attract investment, to ensure that government receives a substantial return from profitable projects, and to encourage efficient mineral exploitation. Their effect is to lighten the tax burden until such time as initial investment and a return on it is recouped; this is particularly so as regards marginal mines, which would enjoy the benefit of accelerated depreciation allowances. It thus facilitates rapid recovery of the initial investment and, in combination with the provision permitting carry-forward of losses in assessment of APT, significantly reduces the risk that application of taxation will result in a mineral investment incurring loss. Both factors are

of course very important from the perspective of a
potential investor. On the other hand, the application of APT to all profits after the investment has
been recovered, the charging of royalties and
dividend withholding tax, and the absence of accelerated depreciation allowances for non-marginal
mines means that profitable operations will quickly
yield a substantial return to government. Yet the
marginal tax rate (70 per cent) is not so high as to
remove the incentive for efficient mineral exploitation, while the low level of royalties (1.25 per
cent of the value of minerals produced) means that
their application is unlikely to significantly
distort production decisions.

While the taxation framework is generally
appropriate, it is also necessary to assess Papua
New Guinea's broader mineral development strategy,
based on the concept of maximising mineral revenues
for use in rural development. In the following
sections the economic and other effects of the
Bougainville mine are analysed, and this should
provide evidence on which to base a judgement.

Generation of Government Revenue

Table 9.1 gives details of the revenue generated by
BCL over the period 1972-80, broken down into its
various components. Income taxes (corporate income
tax, dividend withholding tax) represented by far
the most significant source of revenue, accounting
for K319.7 million[35] or 65 per cent of the total.
Royalties (26.0 million) and Group Tax or Employee
Taxation (K39.5 million) were also important.
Dividend income should be treated separately, since
it results from an investment by government of K25
million. It amounted to K93.6 million, indicating a
net return of K68.6 million.

BCL's contribution to total government revenue
is substantial, though its significance varies
considerably over short time-periods, mainly as a
result of fluctuating metal prices. In 1974/5, for
example, total internal revenue was K180 million,
while receipts from BCL were K96 million in 1974; in
1975/6, revenue was K219 million, while receipts
from BCL were only K32 million in 1975. Over the
period 1972/3 - 1980, revenue generated by BCL's
operations amounted to 26.5 per cent of internally-
generated revenue and 12.1 per cent of total
revenue.[36] Fluctuations in BCL's tax payments do not
have an equivalent effect on funds available to
government, because a portion of revenues is

Table 9.1: Government Revenue Generated by Bougainville Copper Limited, 1970-80* (K million)

| | 1970 | 1971 | 1972 | 1973 | 1974 | 1975 | 1976 | 1977 | 1978 | 1979 | 1980 | TOTAL |
|---|---|---|---|---|---|---|---|---|---|---|---|---|
| Corporate Income Tax | - | - | - | 0.3 | 63.2 | 15.7 | 20.3 | 13.7 | 22.0 | 77.9 | 51.2 | 264.3 |
| Dividend With-holding Tax | - | - | 1.3 | 9.4 | 8.8 | 3.1 | 3.1 | 2.5 | 4.8 | 12.8 | 9.6 | 55.4 |
| Royalties | - | - | 1.0 | 2.9 | 3.8 | 2.2 | 2.6 | 2.5 | 2.8 | 4.0 | 4.2 | 26.0 |
| Group Tax | 0.2 | 0.3 | 1.1 | 1.9 | 2.4 | 3.7 | 5.3 | 5.4 | 5.5 | 6.3 | 7.4 | 39.5 |
| Customs Duties | - | - | 0.2 | 1.1 | 0.9 | 1.9 | 1.6 | 2.3 | 2.2 | 2.8 | 3.0 | 16.0 |
| Dividends | - | - | 2.2 | 16.3 | 14.7 | 5.3 | 5.3 | 4.3 | 8.1 | 21.4 | 16.0 | 93.6 |
| Totals | 0.2 | 0.3 | 5.8 | 31.9 | 93.8 | 31.9 | 38.2 | 30.7 | 45.4 | 125.2 | 91.4 | 494.8 |

* Tax revenue is attributed to the year in which the relevant income was earned.
Source: Information provided by BCL.

retained in a Mineral Resources Stabilization Fund in years of high profit, and utilised when profits are low.[37]

Has BCL engaged in transfer pricing? The Company sells its output on an 'arms length' basis, and so any profit transfer would result from over-charging for goods or services provided by affiliates. CRA provides a number of services to BCL, some directly and some through a wholly-owned subsidiary, MINENCO. A range of managerial, tech-nical, and informational services is provided by CRA itself, for which it receives management fees based on the volume of work undertaken (rather than on a percentage of sales, for example). These fees do not appear excessive given the depth of CRA's involve-ment: from 1974 to 1980, for instance, they averaged less than K1 million per annum.[38]

MINENCO carries out design and construction activities for BCL, on an 'arms length' basis accor-ding to BCL's management, but provincial government officials suspect that MINENCO is used to channel profits out of Papua New Guinea.[39] This is very unlikely. MINENCO's profits are modest, amounting to A$202,000 in 1976 and A$925,000 in 1977.[40] Since MINENCO carries out work for other companies and since it would earn 'legitimate' profits on arms length trading with BCL, any element of hidden profit transfer would be insignificant. It seems certain that BCL/CRA would not risk incurring government displeasure for such paltry returns.

Generation of Employment and Wage Payments

BCL's national workforce increased from 2,594 in 1972 to 3,416 in 1980. Expatriate employment fell from 971 in 1972 to 858 in 1976 as certain skilled positions were taken over by nationals, but has since remained constant at around 850. BCL's oper-ations have also created significant indirect employment on Bougainville. Employment in factory operations increased by over 900 between 1968/9 and 1974/5, while the public sector staff requirement specifically created by the copper project was apparently in excess of 1,000. The majority of BCL's domestic purchases of goods and services are made on Bougainville, and it is consequently unlikely that significant indirect employment has been created outside the region.

The labour employed by BCL has been withdrawn from other uses, usually from agricultural produc-tion, and thus some opportunity costs have been incurred. However, the willingness of the individuals involved to accept wage employment indicates that the change of occupation has enabled them to increase their net incomes.

Wage and salary payments by BCL have totalled K309.4 million over the period 1969-80. A disproportionate share of these payments has accrued to expatriates who have filled the most highly-skilled (and highly-paid) positions; in 1973, for example, the ratio of average expatriate to average national incomes was 6.2. This ratio declined to 3.8 in 1979,[41] reflecting a more rapid growth in average national than in average expatriate earnings.

**Creation of Additional Economic Activity through Linkage Development**

Backward Linkage. BCL purchases only a small proportion of its goods and services in Papua New Guinea, and consequently backward linkage has generally been weak. Local purchases of industrial goods are practically non-existent, for two reasons. First, Papua New Guinea does not produce many of the inputs used by BCL (for example mining equipment, fuel oil, tyres, chemicals). Second, local businesses which might supply BCL are usually structured to sell small quantities of goods at high profit margins on the domestic market. They can rarely match the extremely competitive prices which foreign producers offer BCL on large bulk orders. However BCL believes that individual domestic firms have the potential to compete successfully.[42] The Company's expenditures do support some local service industries, for example motor repairs, electrical equipment repairs, and trucking.

BCL's purchases of agricultural goods have had a major impact on Bougainville. Particularly important are purchases of fruit and vegetables, which averaged some K250,000 per annum in 1976-8 and rose to K440,000 in 1980; in addition, about K100,000 is spent annually on fish and poultry.[43] It has been estimated that vegetable farming provides villagers with incomes which are from 35 to 200 per cent above those from other cash crops, depending on the intensity of farming and the crop used as a basis of comparison.[44] Considerable scope apparently exists for expanding domestic purchases of agricultural goods, for example fish, poultry and particularly beef. BCL does buy beef from mainland areas, but its demand far exceeds the available supply. Some fish is supplied by small-scale operators, but facilities to catch, store and transport fish in large quantities are not available locally.

Clearly some opportunities exist for expanding backward linkage; the obstacles to their development do not arise from any reluctance on BCL's part to purchase locally, but rather from bottlenecks in supply. The Company has contributed substantially to the removal of supply bottlenecks on Bougainville. It set up Panguna Development Pty. Ltd. to assist in the establishment of large-scale enterprises, and formed a Business Advisory Service which has helped set up small-scale trucking, sawmilling and fishing businesses. It also sponsors agricultural extension officers who advise villagers on animal husbandry and cash-crop farming. Action is required on a national basis if opportunities for cattle-raising and fishing are to be fully exploited, and it can only be taken as part of a broader government strategy aimed at enhancing Papua New Guinea's self-sufficiency in food production.[45]

Forward Linkage. Since BCL's exports are entirely in concentrate form, forward linkage is absent. The national government has not yet decided whether it wishes to have mineral processing conducted locally. It is concerned with the environmental impact, suspects that little additional government revenue would be generated, and realises that incomes accruing to nationals would not be substantial because of the high capital intensity of mineral processing and because much of the necessary capital, equipment, operational stores and skilled labour would be imported.[46]

BCL is reluctant to invest in a copper smelter in Papua New Guinea, for a number of reasons. The most important is its belief that such a venture would be unprofitable, mainly because of the existence of surplus capacity in the world copper smelting industry and because of rapidly-rising energy costs.[47] In addition, BCL now enjoys a competitive advantage in marketing its output because of the very high quality of its concentrates, an advantage which would be lost if it sold copper metal. Finally, CRA follows a policy of diversifying its investments across a wide range of minerals, and it might be reluctant to commit additional and very substantial funds to its copper mining activities.

It should be noted that neither CRA nor RTZ possesses copper smelting capacity which relies on BCL as a source of feedstock. BCL's reluctance to construct a smelter is thus apparently not the result of its position as a subsidiary of a multi-national mining corporation, but rather of the application of 'normal' commercial criteria to an investment decision.

Net Foreign Exchange Inflow

BCL has made a major contribution to Papua New Guinea's export income, accounting for 45 per cent of the value of total exports over the period 1972/3 - September 1980. Table 9.2 attempts to calculate the net foreign exchange inflow resulting from its operations. Positive effects result from net sales revenue and from imports of capital, negative effects from a variety of factors. Precise information is available concerning the most important of these (loan repayments, interest and dividend payments, imports of goods and services), while others (expatriate employees' remittances, induced imports) have been estimated on a basis outlined in the notes to Table 9.2. Table 9.2 indicates that net foreign exchange inflow amounted to only 31.0 per cent of gross foreign exchange inflows over the period 1972-80.

In addition, BCL exerted a major influence on the balance of payments during the construction period (1969 - April 1972), when foreign currency to the value of K356 million was expended. The domestic content of local purchases amounted to about K16 million, while K19 million was paid to indigenous employees. Sixty one million kina was paid to expatriate employees,[48] indicating remittances of about K30 million and local expenditures of the same amount (see the notes to Table 9.2). Taxes amounted to K19 million during the main phase of construction (1970-71),[49] indicating total net foreign exchange inflow of K84 million. Gross foreign exchange inflow in 1969-80 was therefore K2,570,790,000, of which it is estimated that K770,678,000 or 30.0 per cent represented net foreign exchange inflow.

In conjunction with Table 9.1 these figures illustrate the central role of payments to government in generating positive balance of payments effects. Over the period 1969-80, government revenue was K513.3 million, equal to 67 per cent of net foreign exchange inflow.[50] Foreign exchange outflows were accounted for by imports (54 per cent), loan repayments (17 per cent), dividends (18 per cent), interest payments (7 per cent) and expatriates' remittances (4 per cent).

Table 9.2: Net Foreign Exchange Inflow from the Bougainville Mine, 1972-80 (K000s)

| | 1972 | 1973 | 1974 | 1975 | 1976 | 1977 | 1978 | 1979 | 1980 | Totals |
|---|---|---|---|---|---|---|---|---|---|---|
| Net Sales Revenue | 95,695 | 249,048 | 279,825 | 184,754 | 205,349 | 200,578 | 223,282 | 339,706 | 335,140 | 2,113,377 |
| Capital Imports* | 36,300 | 14,946 | - | - | 13,120 | 28,047 | 5,800 | 3,200 | - | 101,413 |
| Total Inflow(A) | 131,995 | 263,994 | 279,825 | 184,754 | 218,469 | 228,625 | 229,082 | 342,906 | 335,140 | 2,214,790 |
| Debt Repayment | - | 68,002 | 36,904 | 18,462 | 19,544 | 25,635 | 64,642 | 21,041 | 11,625 | 265,855 |
| Interest Charges | 22,562 | 17,493 | 13,205 | 11,469 | 11,060 | 11,661 | 6,093 | 4,538 | 3,546 | 101,627 |
| Dividends Paid Overseas | 7,359 | 54,453 | 49,187 | 17,887 | 17,887 | 14,309 | 26,831 | 53,663 | 40,254 | 281,830 |
| Expatriates' Remittances** | 2,500 | 4,800 | 6,000 | 6,400 | 7,400 | 7,800 | 7,600 | 8,000 | 9,000 | 59,500 |
| Imported Goods and Services+ | 60,400 | 52,700 | 56,300 | 81,600 | 98,600 | 103,600 | 104,600 | 116,300 | 145,200 | 819,300 |
| Total Outflow (B) | 92,821 | 197,448 | 161,596 | 135,818 | 154,491 | 163,005 | 209,766 | 203,542 | 209,625 | 1,528,112 |
| Net Foreign Exchange Inflow = (A)-(B) | 39,174 | 66,546 | 118,229 | 48,936 | 63,978 | 65,620 | 19,316++ | 139,364 | 125,515 | 686,678 |
| - As % of (A) | 29.7 | 25.2 | 42.3 | 26.5 | 29.3 | 28.7 | 8.4++ | 40.6 | 37.5 | 31.0 |

* Capital imports are attributed to the year in which the relevant foreign currency loans were drawn down.
** It is not known precisely what proportion of expatriates' wages is remitted; BCL estimates that 45 per cent of all expatriate income is remitted.
+ Includes import content of domestic purchases of goods and services, estimated by BCL to be 50 per cent.
++ The very low figures for 1978 reflect BCL's decision to make substantial loan repayments from accumulated reserves.
Source: Table 9.1; information provided by BCL.

## Provision of Scarce Capital Resources

In the context of the domestic economy, the capital requirements of the Panguna project have been enormous. This is illustrated by the fact that BCL's capital expenditure amounted to nearly twice the sum of those by all other private concerns and by government over the period 1969-72.[51] Finance had to be sought abroad, and as mentioned above RTZ's knowledge of and links with the international banking system played an important part in ensuring that it would be forthcoming.

## Provision of Technology, and Technical and Other Skills

The technology utilised in exploiting low-grade copper deposits of the Panguna type was well established and could probably have been obtained without direct foreign investment. However, Papua New Guinea was almost totally lacking in the skills required to apply that technology in developing and operating a major mining project. It also lacked the entrepreneurial expertise to marshall the other resources needed for large-scale mineral development.

BCL has filled its requirements for skilled personnel by recruiting expatriates and by establishing a comprehensive training programme for its national employees. BCL's role in ensuring the availability of skilled expatriates is especially important at the senior technical and executive levels, since it will be some time before suitably-qualified nationals are available to fill these positions. It is difficult for companies operating in LDCs to attract and retain highly-qualified expatriates, but the efficiency of BCL's operations has not been affected by this problem. The Company's position as part of a large international mining organisation is probably important in this regard – prospective employees know that they would enjoy security of employment if operations ceased at Panguna or if they found conditions on Bougainville unacceptable.

BCL's training programme is based on an apprenticeship scheme and on a scholarship programme which funds students attending UPNG and other colleges. The programme causes an impact beyond the Panguna project itself. Turnover among employees is high (trainees do not bond themselves to work for BCL), and the Company must train more workers than it will employ at any one time. Consequently the supply of skilled labour to the rest of the economy is enhanced. Over the period 1968-80, 126 BCL-sponsored university students graduated and of these at least 38 either did not join BCL or have since terminated their employment. During the years 1970-80, over 180 skilled tradesmen (mainly mechanics, welders, heavy equipment fitters, plumbers and carpenters) left the Company;[52] to place this figure in perspective, Papua New Guinea's total annual output of skilled tradesmen was about 300 in the early 1970s. Substantial numbers of other skilled and of semi-skilled workers have also terminated their employment.

Not all of those who leave BCL remain in the wage-earning workforce; a proportion return to their villages to open trade stores or develop family land, and some of these do not utilise the skills they have acquired. On the other hand certain trades (motor mechanic, carpenter, fitter) could be useful in the conduct of small businesses and in village development projects. Some skilled workers do remain in the wage-earning workforce, as apparently do nearly all graduates, and since both categories are in short supply in Papua New Guinea, BCL's contribution is clearly a valuable one.

BCL's training effort has generally attracted favourable comment,[53] but some company officials involved in training feel that it is unsatisfactory, particularly in the very important area of on-the-job training. In their view, expatriate staff lack any interest in passing on their skills since they would be 'training themselves out of a job'. They claim that BCL is partly responsible for this situation because it fails to systematically monitor the progress of apprentices to ensure that skills are in fact being imparted, and they argue that localisation will be slow until the problem is overcome.[54] This argument seems to find support in the absence of any decline in expatriate employment over the period 1976-80. Vigorous action by the Company might well cause dissatisfaction among expatriate staff, and BCL's alleged lack of initiative might reflect its unwillingness to risk a confrontation.

It is unlikely that its foreign nature is responsible for any such attitude on BCL's part. The Company's primary concern is to maintain efficiency, and at present efficiency depends on a high degree of commitment from expatriate workers. BCL thus faces a dilemma. While expatriates are needed, issue will not be taken with them on the question of training, but until training is improved expatriates will be needed. This dilemma faces other mining

industries which depend on expatriate labour, and it
is a dilemma which continued to face Zambia, for
example, after foreign mining companies had been
nationalised. Its basic cause lies in the character
of the national workforce when mining commences.

Provision of Infrastructure

The infrastructure provided by BCL has had significant 'spin-off' effects, particularly by facilitating movement of goods into and out of the region.
By giving villagers in the vicinity of the mine and
in Southwest Bougainville access to the port of
Kieta, the construction of the Panguna-Kieta road
has substantially lowered transport costs for export
crops and for inwards shipment of general cargoes.
Local people have increased their incomes, and will
do so again at a later date since these improvements
have encouraged additional planting of tree-crops.
     Part of mine infrastructure has been provided
by the government. Detailed information is not
available regarding the cost of providing services
to BCL or the charges made for them, but the
government apparently suffers a slight net loss from
their provision.[55]

Absorption of Domestic Resources

BCL's operations have absorbed significant amounts
of agricultural land and of labour on Bougainville.
Cash crop production and subsistence farming have
been adversely affected by the resumption of the
Arawa plantation and of other areas, and by withdrawal of labour.[56] There is no agreement as to what
the effects of labour withdrawal will be in the
longer term. Some claim that a permanent loss of
agricultural skills, and so of production, will
occur, but it is possible that areas now abandoned
or under-utilised will again be fully exploited as
Bougainville's population grows.

Impact on Consumption Patterns

BCL's presence has influenced consumption patterns
on Bougainville in a number of ways. Villagers in
the vicinity of the mine must now purchase much of
their food requirements, previously met by subsistence farming. A substantial number of people have
obtained large cash incomes from compensation payments or wages, enabling them to greatly extend the
range of items consumed. Many Bougainvillians have

been influenced by their experiences in company messes, canteens and supermarkets, and by the 'demonstration effect' exercised by expatriate workers and their families. They feel new and different needs and have changed their consumption patterns in satisfying them.

But in what sense, if any, are the new patterns of consumption inappropriate? First, they add to Papua New Guinea's imports (for example of tinned and fresh food, motor vehicles, luxury goods), not a welcome development for a country which emphasises the need for self-sufficiency. Second, consumption of certain items, particularly alcohol, is having socially disruptive effects. Third, Bougainvillians may not be able to meet their newly-acquired needs in the future. Many of those who have received compensation payments from BCL will not enjoy a regular income, and there is a clear danger that these payments, made to allow replacement of income-generating assets (land, forest, rivers), will be frittered away on consumer goods. More broadly, the Kieta district has become highly dependent on the mine and, if and when mining ceases, a serious regional problem may emerge as few alternative sources of wage employment are likely to be available. This difficulty could be overcome if wage payments and other current cash receipts were invested in assets which could provide incomes in the future, but the impact of the mine on consumption patterns reduces the likelihood of this happening. In a detailed survey carried out in 1973-4, Moulik found that villagers in the Kieta area had substantially higher consumption levels and lower savings levels than those in the Buin district, farther removed from the mine.[57] Ironically, the Buin people have not lost their income-generating assets, and consequently have less need to save.

Impact on Domestic Income Distribution

BCL's demand for labour has created upward pressure on wage rates for plantation workers on Bougainville, and so has redistributed income from expatriate planters to their indigenous employees, a desirable outcome given the Somare government's policy priorities. However in another way BCL's presence skewed income distribution in favour of non-nationals, since it brought large numbers of highly-paid expatriates to the country. The Company's demand for labour has been less significant on the national level; it may have exercised some upward pressure on general wage levels, but it would not be possible to identify the distributional effects of any such trend. It is unlikely that payment of high wages to mineworkers has exercised a significant 'demonstration effect', since the numbers involved are small (less than 0.9 per cent of wage earners) and since BCL's operations are geographically isolated. However the high profits generated by BCL in 1973-4 were apparently an important factor behind successful political pressure for large increases in minimum wages and public service salaries in 1974.

BCL's operations have had a marked effect on regional income distribution. In 1966/7, domestic factor incomes per capita on Bougainville exceeded the average for the rest of Papua New Guinea by an estimated 9 per cent; by 1971/2, this figure had risen to 185 per cent.[58] Since a high proportion of Bougainvillians living in rural areas remained heavily dependent on subsistence farming during these years, income accruing within the region was apparently much more unequally distributed at the end of this period than at the beginning.

Mineral Policy - a Broader Perspective

The strategy of maximising mineral revenues for use in rural development is basic to Papua New Guinea's general mineral policy. This analysis of the Bougainville mine's economic impact indicates that the assumptions which underlie this strategy are correct. The direct economic impact is slight; employment is modest, linkage industries practically non-existent (with the exception of food crop cultivation on Bougainville), and the spin-off effects of mine infrastructure are highly localised. The direct impact which occurs does have undesirable features, particularly in that it encourages inappropriate consumption patterns and adversely affects regional income distribution. On the other hand, the Bougainville mine has generated substantial economic rents, a significant proportion of which have been captured by the national government. Thus maximisation of mineral revenues does appear to provide the most logical response.

A policy of applying mineral revenues to rural development is clearly appropriate for a country whose population is predominantly rural and will be so for the foreseeable future, and in which allocation of resources has previously been heavily weighted in favour of the urban sector. However one important question does arise: can the government ensure that mineral revenues actually reach the rural sector?

In the mid 1970s there were indications that a high proportion of such revenues might rather be absorbed by urban-based civil servants and on urban-oriented services. Over the period 1971-5, for example, government expenditure on salaries and wages rose by 83 per cent, an increase due to expansion of the public service from 37,000 to 50,000 and a rise in average public service earnings of 35 per cent.[59] It is more difficult to trace developments in the allocation of expenditure on services between urban and rural areas, but over the years 1972-5 Departmental expenditure (much of which would be absorbed in maintaining the existing urban-oriented structure of services) increased by 212 per cent, while capital expenditure (which would include items crucial to rural development such as construction of roads, marketing and storage facilities, schools and hospitals) fell by 22 per cent.[60]

Since 1976, the government has tried to come to grips with these problems. It is attempting to limit growth of the public service and has introduced an incomes policy designed to prevent any real increase in public service incomes. It has also attempted to curb expenditure on existing services and infrastructure and has increased charges for government services, so that expenditure can be re-oriented towards capital investments in rural areas. However, the government has enjoyed only limited success in implementing these policies as is made clear, for example, by the Bank of Papua New Guinea's annual report for 1981; indeed in some respects, the situation appears to have worsened.[61]

It is important to note that the national government's failure to apply mineral revenues to rural sector development is apparently not due either to the origin of those revenues or to any effect of BCL's operations. This is evident from the record of the North Solomons Provincial Government (NSPG), which obtains about a third of its revenues from mineral royalties and administers an area which feels the influence of BCL's operations much more strongly than Papua New Guinea as a whole. The NSPG has achieved substantial success in channeling revenue into rural development projects; for example, over the period 1979-81 over 70 per cent of its total expenditure has been absorbed by capital works, a high proportion of which involve construction of roads and bridges designed to facilitate increased economic activity and improved access to services in rural areas of Bougainville and Buka.[62]

Notes

1. PNG House of Assembly Debates, 16 November 1972.
2. Ibid.
3. Ibid.
4. IBRD, Economic Development of Papua New Guinea, pp. 34-6.
5. Government Printer, Port Moresby, September 1968.
6. Ibid., p. 104.
7. P. Hastings, New Guinea: Problems and Prospects (Cheshire, Melbourne, 1973), p. 83.
8. R.T. Shand and M.L. Treadgold, The Economy of Papua New Guinea: Projections and Policy Issues (Department of Economics, Research School of Pacific Studies, Australian National University, Canberra, 1971), pp.54, 71.
9. M.L. Treadgold, Economic Growth and Inequality in Papua New Guinea (The University of New England, 1976), p. 10.
10. Faber Report, p. 6.
11. Treadgold, Economic Growth and Inequality p. 12.
12. Faber Report, p.4.
13. Ibid., pp. 27, 101.
14. In 1971, BCPL became Bougainville Copper Ltd. (BCL).
15. R. Garnaut, 'The Framework of Economic Policy-Making', in J. Ballard (ed.), Policy-Making in a New State: Papua New Guinea 1972-77 (University of Queensland Press, St Lucia, 1981), p.194.
16. R. Garnaut and A. Clunies Ross, 'Uncertainty, Risk Aversion and the Taxing of Natural Resource Projects', Economic Journal, Vol. 85 (June 1975), pp.272-87.
17. Ibid., p. 277.
18. Interview with Department of Finance official, Port Moresby, 8 June 1978.
19. Ibid. The discussion of government policy is based on this interview and on interviews with the following: Assistant Secretary, Office of Minerals and Energy, Port Moresby, 27 June 1978; Assistant Director, Central Planning Office, Port Moresby, 27 June 1978; former Secretary to the PNG Cabinet, Port Moresby, 11 June 1978.
20. Nationalisation without compensation was not considered, partly because the government did not believe that it possessed the administrative or technical skills to run a major mining concern, partly because of the adverse effects such action

was likely to have on the flow of investment and aid to Papua New Guinea.
21. Any increase in BCL's costs would of course reduce profits and government revenues.
22. This was certainly how its attitude was interpreted by the government. Interview with Ross Garnaut, Canberra, 23 October 1979.
23. Ibid.
24. *Mining (Bougainville Copper Agreement) (Amendment) Act 1974* (Government Printer, Port Moresby, January 1975), Clause 16A.
25. Ibid., Clause 16.
26. Interview with Ross Garnaut, Canberra, 23 October 1979.
27. Confidential Source.
28. Interview with Ross Garnaut, Canberra, 23 October 1978.
29. Company negotiators had claimed there were certain 'principles' on which it could not compromise, for example non-retrospectivity of tax liability.
30. *Mining (Bougainville Copper Agreement) (Amendment) Act*, Clauses 7i, 7j, 16b, 26a.
31. Interview with former Secretary to the PNG Cabinet, Port Moresby, 11 June 1978; interview with former Political Adviser to the Minister for Finance, Port Moresby, 8 June 1978.
32. This is not to suggest that Papua New Guineans did not suffer racial discrimination under the colonial regime; but very few, if any, of Papua New Guinea's leaders experienced the sort of highly-organised, institutionalised and all-pervading discrimination suffered, for instance, by their Zambian counterparts.
33. Papua New Guinea did not then have a written constitution, and was governed by ordinances and orders-in-council which could be amended by Parliament.
34. For a detailed outline and discussion of the tax regime, see PNG, 'Financial Policies Relating to Mining and Mining Taxation: Statement of Intent' (Port Moresby, June 1977); O'Faircheallaigh, 'Foreign Investment in Mineral Development', pp. 466-73.
35. A new currency, the Kina (K) was established in 1975; initially one Kina equalled one Australian dollar.
36. C. O'Faircheallaigh, 'The Economic Impact of the Bougainville Copper Project', Paper presented to the Waigani Seminar on 'Trade, Investment and Development in the Pacific', University of Papua New

Guinea, Port Moresby, 6-10 September, 1981.
37. For an account of how the Fund operates, see N.V. Lam, Fiscal responses to export instability in Papua New Guinea, IASER Discussion Paper No. 19 (Port Moresby, April 1978), pp. 12-16.
38. Information supplied by BCL.
39. Interview with provincial government offical, Arawa, 20 June 1978.
40. CRA, Annual Report, 1977, p. 50.
41. Derived from information provided by BCL.
42. Two specific areas mentioned were timber and wood products, and printing. Interview with BCL's Chief Purchasing Officer, Panguna, 15 June 1978.
43. Information supplied by BCL.
44. W.D. Scott and Company Pty. Ltd., 'A Study of the Impact of the Bougainville Copper Project on the Economy and Society of Papua New Guinea', unpublished, Sydney, January 1973, 3/39.
45. Food represents a major import item for PNG, amounting to 20 per cent of total imports in 1975.
46. Interview with Assistant Secretary, Office of Minerals and Energy, Port Moresby, 27 June 1978.
47. Interview with BCL's Executive Manager, Commercial, Panguna, 26 June 1981.
48. These figures were obtained from Mikesell, Foreign Investment in Copper Mining, p. 102.
49. W.D. Scott and Company Pty. Ltd., 'Impact of the Bougainville Copper Project', Appendix 5A-19.
50. Derived from Tables 9.1 and 9.2.
51. PNG Bureau of Statistics, Capital Expenditure by Private Businesses, various issues; and Hastings, New Guinea, Appendix, Table v.
52. Information supplied by BCL.
53. See, for example, Mikesell, Foreign Investment in Copper Mining, p.118.
54. Confidential Source.
55. Interview with Ross Garnaut, Canberra, 23 October 1979.
56. For a detailed analysis of the impact of labour withdrawal on village occupations and cash cropping, see T.K. Moulik, Bougainville in Transition, Development Studies Centre Monograph No. 7 (The Australian National University, Canberra, 1977), pp. 36-8.
57. Moulik, Bougainville in Transition, pp. 62-3, 166-7.
58. Derived from Treadgold, The Regional Economy of Bougainville, Table 8, p. 35; Table 11, p. 41.
59. Derived from World Bank, Papua New Guinea: Its Economic Situation and Prospects for Development (Washington, 1978), Table 6.4, p. 103, p. 104.

60. Derived from Ibid., Table 6.4, p.103.
61. Bank of Papua New Guinea, Annual Report, 1981, pp. 39, 44.
62. North Solomons Provincial Government, North Solomons Provincial Budget, 1979, 1980, 1981, Arawa.

Chapter Ten

THE IMPACT OF FOREIGN-FINANCED MINERAL DEVELOPMENT

The case study material allows certain conclusions
concerning the effects of foreign-financed mineral
development and the nature of host country-foreign
investor relations. I begin by summarising the
empirical evidence yielded by the case studies,
after which the implications of the evidence for the
nature of the host country-foreign investor relationship are considered in the light of the
discussion in Chapter One.

Creation of National Incomes

In all four cases, substantial profits were
generated by foreign mining operations. The host
countries differed considerably in the extent to
which they captured those profits, and to which they
reconciled their need for revenue with the need for
continued exploration and development.

Ireland. Prior to 1973, Ireland sacrificed revenue
unnecessarily in its attempts to attract foreign
mining investment. Its failure to capture a higher
proportion of economic rents occurred despite
declining uncertainty as to geological and other
conditions in Ireland and an increase in the expertise available to government. In attempting to overcome Ireland's general economic problems, Irish
governments offered generous tax concessions to
encourage foreign investment. The attitude of Irish
politicians towards foreign mining companies led
them to accept claims that similar concessions were
necessary to attract foreign mining investment. Some
of the companies involved apparently played a part
in fostering those attitudes.
    The discovery of the Navan zinc/lead deposit in

The Impact of Foreign-Financed Mineral Development

1970 substantially increased the Irish government's
'objective' bargaining power, and led to public
pressure for termination of tax concessions to
foreign mining companies. Those concessions were
withdrawn, but bureaucratic decision-makers were
able to negate substantially the effect of their
withdrawal. The Irish government's total 'take' from
new mineral developments is also influenced by the
level of its equity participation and by the level
of royalties, both negotiated with the companies
involved. In the negotiations with Tara Mines
Limited, constitutional, political and bureaucratic
constraints unconnected to the host country-foreign
investor relationship played a crucial role in
weakening the government's bargaining position.

Zambia. In the mid and late 1960s, the Zambian
government captured a very high proportion of mining
company profits, reducing the expected rate of
return below that required to attract new
investment. That rate was certainly high, reflecting
the degree of risk attached by foreign mining
companies to investment in Zambia. However, the
government could have acted to reduce risk perceptions without sacrificing substantial amounts of
revenue, particularly by changing the manner in
which taxation was applied. This was done in 1970,
but the new tax rate is substantially higher than
the average over the period 1964-70 and, in
conjunction with the requirement that the government
have a 51 per cent 'carried' interest in all
exploration ventures, its application very probably
deters, and will deter, exploration for and development of base metal deposits.

Thus in attempting to meet its revenue requirements, Zambia has paid insufficient attention to the
need for continued investment. To some extent, this
failure reflected the urgency of those requirements
and the absence of alternative ways of meeting them.
It also reflected the influence of historical and
political factors which militated against rational
decision-making on issues involving foreign mining
investment.

Australia. During the period to 1972, Australian
governments unnecessarily sacrificed revenue in
their attempts to encourage foreign mining
investment. This situation resulted from a number of
factors. First, Australian authorities were not
fully aware of the factors favouring Australia as a
mineral investment location in the post-war period.

This lack of awareness apparently reflected ignorance rather than any element of 'complicity': the factors involved were complex, varied, and occurred in different parts of the world, and consequently their implications for Australia were not grasped by policy-makers. As a result, Australian authorities believed that mineral development would not occur if tax concessions were not granted. Second, Australian governments over-estimated the likely extent of direct economic benefits from mineral development. The sacrifice of revenue consequently seemed worthwhile, and the direct benefits seemed sufficient reward for permitting exploitation of Australia's minerals.

Papua New Guinea. Papua New Guinea's colonial administrators failed to negotiate appropriate taxation arrangements for the Bougainville Copper project by omitting to provide for an unexpectedly favourable outcome. In this instance, lack of relevant expertise was an important factor, though it is possible that Australian administrators felt a community of interests with the foreign investors concerned, resulting in a failure to protect fully their own real interests and those of Papua New Guinea.

Papua New Guinea's national governments have enjoyed considerable success in reconciling their revenue requirements with the need for continued investment and for efficient mineral exploitation. Substantial revenues were obtained as a result of the Bougainville renegotiation, while an efficient, generally-applicable mineral taxation system has been designed which will obtain substantial revenues from profitable projects, but which is unlikely to deter mining investment or encourage inefficient working of minerals. Papua New Guinea's success in this regard resulted from the high degree of rationality which characterised decision-making, itself reflecting historical and political factors, particularly the absence of strong racial antagonism and of domestic political or bureaucratic groups with an interest in influencing the outcome.

Generation of Employment. In all cases, mining employment was small in relation to the value of output, reflecting the capital intensity of mining operations. No evidence was found that the mineral deposits involved could have been exploited in qualitatively different ways, and only in Zambia were economically-feasible mining methods suffic-

iently flexible to allow significant cross-substit-
ution between labour and capital. In this case it
was argued that the profit-maximising factor
combinations pursued by foreign mining companies
were not necessarily inimical to Zambia's interests,
but the Zambian government's failure to establish
relevant policy priorities renders a more definite
judgement difficult.

In all four cases, wage payments to mineworkers
were substantially above the national average,
reflecting the high productivity of labour in a
capital-intensive industry. In Papua New Guinea and
Zambia very large wage differentials existed between
expatriate and national mineworkers, and in the
Zambian case those differentials exercised a major
impact on wage levels in the monetary sector and on
domestic income distribution (see below).

Backward Linkage. Substantial economies of scale are
associated with production of many of the inputs
utilised in modern mining operations. In addition,
the total scale of backward linkage from mineral
development is comparatively weak, as the Australian
case illustrated. Consequently, mining operations
can rarely of themselves provide the basis for
establishment of new efficient manufacturing industries. Linkage usually occurs where the relevant
industries are already in existence, or where a high
proportion of the fixed overhead costs associated
with production of mining inputs can be met from
established operations. In other words, the existing
level of industrialisation in the host country is a
crucial factor in determining the extent of linkage
development. This is evidenced by the success of
Australia (which also derived important economies of
scale from the large size of its mining industry)
and Ireland in taking advantage of the markets
provided by mineral development. Papua New Guinea
has been unable to do so, and while Zambia has
established a wide range of supplying industries
some have not achieved minimum economies of scale.
It is not evident that the establishment of inefficient manufacturing industries will enhance the
prospects for broadly-based economic development in
Zambia.

The foreign mining companies encountered in the
case studies did display a tendency to purchase
goods and services from suppliers of proven reliab-
ility, which frequently meant suppliers from their
home countries. This led to conflict with host
country authorities in Zambia, but for ideological

and political rather than economic reasons: the Zambian government's aim of disengagement from Southern Africa was incompatible with continued reliance on suppliers in that area. In both Ireland and Australia, this tendency on the part of foreign mining companies was partly responsible for an initial lag in linkage development, which was, however, quickly overcome. Indeed at least some foreign companies operating in these countries have adopted a policy of favouring domestic producers. Their willingness to do so reflects the ability of these producers to supply goods and services at competitive prices, but it should be noted that the Zambian companies also were generally willing to utilise domestic sources of goods and services where these were available, despite the cost disadvantages this sometimes entailed.

It is important to remember that the companies encountered in this study generally did not themselves produce mining inputs. In addition, many of the subsidiaries were responsible for their own purchasing and did not utilise centralised purchasing agencies. Bougainville Copper Limited, for example, itself organises a comprehensive research programme aimed at identifying the most competitive source of the major inputs it utilises, and purchases goods and services accordingly.[1]

Forward Linkage. In each case study country situations were encountered in which foreign investors displayed a reluctance to undertake mineral processing. In Australia, Papua New Guinea and Zambia significant economic constraints militated against local processing and/or fabrication, but it is difficult to judge the extent to which their existence explained the reluctance to invest. In Zambia and Papua New Guinea perceptions of political risk and/or the corporate strategies of the investors concerned were certainly important. In neither case was clear evidence found that these factors resulted in failure to develop viable opportunities for forward linkage, though the Zambian government had to act to ensure that an opportunity for domestic copper fabrication was grasped. In Ireland, at least one major foreign mining company was willing and indeed anxious to invest in processing, but due to factors beyond its control plans to do so did not come to fruition. It should be noted that Ireland enjoyed considerable economic advantage as a location for processing, being close to major European markets and inside the EEC's tariff barriers.

## Provision of Scarce Resources through the Investment Process

Foreign investors have played a crucial role in supplying scarce risk capital in each of the case study countries. This continued to be the case in Zambia even after nationalisation; most of the capital involved was in loan form, but foreign equity capital has financed a high proportion of exploration expenditure since 1970, and such capital will probably be used in future major base metal developments.

In most cases the technology utilised in mineral development was widely available and could probably have been obtained without foreign direct investment; a notable exception involved technology associated with bauxite processing in Australia. However, in each case study country foreign investors played a crucial role in providing the skills required to apply technology efficiently. In each case they also provided the entrepreneurial and organisational skills needed to assemble the other resources required for successful mineral development.

In general little evidence was found that significant opportunities existed for varying the type of technology employed in mineral development. The only exception involved limited opportunities for varying the labour intensity of technology utilised in Zambia's copper mines, but as mentioned earlier the absence of clear policy priorities in this area makes it difficult to assess the appropriateness of the technology employed.

## Creation of External Economies

The specialised nature of the processes involved, the secret nature of some of them and the general absence of domestic mining companies served to reduce the significance of technology 'spill-over' in the host countries.

As regards diffusion of skills, the experience of the developing countries differed substantially from that of Ireland and Australia. In the latter diffusion occurred rapidly, with nationals replacing expatriates over a short period of time. Zambia and Papua New Guinea encountered major difficulties in the area of localisation, mainly because of a shortage of nationals with the education or training to receive skills and because of the unwillingness of some expatriate workers to pass on their expertise. As a result, Zambia and Papua New Guinea have had to attract and retain large numbers of expatriates over extended periods of time. The links between foreign subsidiaries and their parent companies played an important role in allowing them to do this. Zambia has recently attempted to hire skilled personnel independently of the foreign companies, but has encountered serious difficulties in meeting its requirements, particularly at the middle and senior technical and managerial levels. Thus the difficulties encountered in localisation both reduce the contribution of the foreign investor to the host country's store of skilled personnel and prolong the foreign investor's role as a supplier of such personnel.

In general, mine infrastructure did not facilitate emergence of new industries in the host countries. Thus its impact was largely determined by existing patterns of economic activity in mining areas, and its principal effect was to lower costs (particularly transport costs) or improve market access for established enterprises. The geographical location of mineral deposits in Australia, Papua New Guinea and Zambia meant that its impact in this regard was highly localised.

## Creation of External Diseconomies

The operations of foreign mining companies and the presence of their expatriate staff affected consumption patterns in Papua New Guinea and Zambia in ways which imposed costs on those countries, or which threaten to do so. The remote situation of the Panguna mine has helped to localise its impact in this regard, but its effects on Bougainville itself have been substantial. In Zambia, the impact has been more widespread, reflecting the scale of mining activity, its central geographical location, and the length of time over which it has occurred.

Domestic resources were absorbed in each of the case study countries but the implications of resource absorption differed from case to case. In certain instances the resources had previously been idle (labour in Ireland). In others resources were apparently used less efficiently in mining than they would have been in alternative uses (domestic equity capital mobilised by foreign mining companies in Australia). In Papua New Guinea and Zambia agricultural land and labour were absorbed, resulting in significant opportunity costs and in economic and social disruption on a local or regional basis. However the individuals involved were given an opportunity to enhance their incomes, while from a national perspective resources were probably used more productively by the foreign investors than they otherwise would have been.

Foreign mining investment apparently did not exercise a significant effect on domestic income distribution in Australia or Ireland. It had a substantial impact on regional income distribution in Papua New Guinea, widening disparities in wealth between Bougainville and other parts of the country and within Bougainville itself. In Zambia, its effects were severe and deep-seated. Zambian mine-workers received large wage increases over the period 1964-6. The direct effect of these increases was to redistribute income from foreign mining companies and the Zambian government to the mine-workers; indirectly they reduced Zambia's capacity to expand urban employment, widened income dispar-ities between the rural and urban sectors, and lessened the government's ability to redistribute income in favour of the former. This in turn under-mined the prospects for rural development. The situation was exacerbated by the Zambian government's failure to spell out and take account of the implications of its mineral policies for domestic income distribution.

## The Political Impact

Questions of government control over corporate decision-making played a central role in host country-foreign investor relations in the case study countries, with the exception of Papua New Guinea. It is possible to identify three distinct areas in which government sought to influence decisions or exercise control. The first involved what might be called 'strategic' decisions, relating, for example, to disposal of mine output, production and investment policies. The second concerned operat-ional decisions of a financial kind, and particul-arly pricing policies. A third area related to the actual conduct of mining operations, and involved issues such as grading and dilution policies. Issues relating to national economic management were generally not significant. Foreign-financed mineral development did create or help to create problems of economic management (for example through its impact on wage levels in Zambia and on the balance of payments in Australia), but these outcomes were not perceived by host country authorities as resulting from avoidable actions by the companies concerned.

The question of control over production and investment decisions figured most prominently in Zambia, mainly because of factors associated with the structure of its mining industry. In Australia, Ireland and Papua New Guinea, many of the investment decisions taken during the 1950s and 1960s were by companies contemplating investment in the country concerned for the first time. If the decision was negative, there was consequently little basis on which host country authorities could attempt to exercise influence. In Zambia, investment decisions were mainly being taken by operators of existing mines, which gave the Zambian government significant scope to exert influence.

In addition, Australia, Ireland and Papua New Guinea followed a policy of encouraging favourable investment decisions, usually by granting tax concessions. Since there were then few profitable mines operating in these countries, little immediate sacrifice of revenue was involved and, since policy-makers generally believed that mineral development would not occur without the concessions, any future sacrifice was perceived as hypothetical. In Zambia, the companies concerned were generating substantial profits, and tax concessions would represent an immediate and real sacrifice of revenue.

Finally, many of the companies operating in Australia, Ireland and Papua New Guinea were increasing their market shares or maintaining their shares of rapidly-expanding markets. Considerable expansion of capacity was consequently occurring, which seemed to justify the tax concessions, concessions which in turn encouraged further expansion (a point made in relation to Austrlaia by the Fitzgerald Report). The companies in Zambia, on the other hand, were apparently content with their market shares and indeed were willing to see those shares decline.[2]

It is not possible to state conclusively why AAC and RST were not attempting to expand output in Zambia. Changes in relative production costs and the desire to diversify the geographical spread of its investment may have motivated AAC, which did expand its copper interests in Botswana, Canada and Mauritania during the 1960s.[3] RST did not follow a similar course, but it was in a weaker position to do so,[4] and it may be significant that the expansion of Zambian capacity which did occur over the period

1964-9 was entirely due to RST.[5] Concern with over-supply in the copper market was certainly an important factor for both companies - they were earning a comfortable return at prevailing prices, and feared that increase production would lead to a decline in prices. Both also perceived Zambia's political situation as highly unstable.

The Zambian experience seems to support Norman Girvan's contention that the investment and production decisions of the foreign investor will eventually come into conflict with the host country's interests. However, three points should be made here. First, it is not obvious that a strategy of restricting production in order to maintain prices was necessarily contrary to Zambia's interests. If its intention was to use mineral revenues to finance rural and other non-mining development, price maintenance might be a more important consideration than achieving increased production, and so higher employment and export receipts.[6] Two factors explain the Zambian government's unwillingness to consider such a strategy. First, it was aware that it alone could not effectively maintain prices, and believed that other copper producers would expand output regardless. While this belief was probably well-founded, it does not alter the fact that corporate and national interests may not have been in conflict. Second, it had not established clear priorities in relation to the role which mineral development was expected to play, and consequently had not decided whether its priority was price-maintenance or expansion of production. Indeed, it had not even recognised that the two might be incompatible.

A second point is that corporate behaviour was partly determined by the Zambian government's policies. Whatever the reason for the reluctance of AAC and RST to invest in new capacity, the actions of the Zambian government certainly reinforced it, as is evident from the fact that planned capacity expansions were abandoned after the failure to resolve the royalty issue. Third, a host country need not forgo substantial amounts of revenue from profitable mining projects in devising taxation arrangements likely to encourage favourable investment decisions, as the Papua New Guinea case demonstrated.

The Zambian government believed that production would not expand until it established control over the copper industry, which it attempted to do through the 1969 Take-over. This action added substantially to the reluctance of AAC and RST to invest in Zambia, while the arrangements it negotiated did not give the government the control it sought. This failure was due to the timing of the Take-over, which reflected short-term political considerations, and to historical and institutional factors which affected the quality of advice the government received and the quality of decision-making generally. It led to the redemption of the ZIMCO bonds in 1973, but this initiative did not give the government full control, because it failed to carry through the strategy for eliminating the minority shareholders' monopoly of technical expertise. Once again, the reasons for this failure involved political processes unrelated to the host country-foreign investor relationship.

The attempts to gain control of the copper industry imposed substantial costs on Zambia, both in financial terms and as a result of the loss of technical and managerial skills which accompanied the partial withdrawal of foreign investment.

Pricing policies were an important issue in Ireland, Australia and Zambia, though it was argued that, with the exception of a small number of companies in Australia, the activities of the foreign investors involved did not in fact warrant concern. The situation in Ireland was exceptional in that the government was concerned primarily with the pricing policies of multinational oil companies, and felt it had to establish as a general principle its right of access to relevant corporate information. Australia's Labor government over-estimated the degree to which pricing policies inimical to Australia's interests were being practised, though some aluminium companies certainly were engaged in transfer pricing, and fragmentation among mineral exporters did sometimes exacerbate the impact of adverse market conditions on prices.

Concern with obtaining control over the technical aspects of mining operations was largely confined to Zambia. The Zambian government attempted to break the foreign investors' monopoly of technical expertise by establishing monitoring units designed to implement selective checks on mining operations and on 'technical' aspects of the decision-making process, but it followed this course only briefly. The reasons for its abandonment related primarily to political processes unconnected to the host country-foreign investor relationship.

It should be noted that the _need_ for host country control over mining operations is influenced by government policies. In particular, taxation structures can have an important bearing in that they can either encourage or discourage efficient mineral exploitation. Thus, for example, the application by the Zambian government of substantial unit royalties to copper production encouraged the high-grading perceived as one factor which rendered government intervention necessary.

In general, neither foreign mining companies nor their home countries attempted to subvert political processes in the host countries examined; only in Ireland was evidence found that foreign mining companies had influenced these processes so as to produce outcomes favourable to themselves.

Three general points emerge from this summary. First, foreign-owned mining companies are not all alike, and the differences between them have very important implications for the way in which foreign mining investment affects a host country. The work of Moran and Girvan, for example, is based on a very specific model of corporate behaviour. In this model, the typical company is vertically integrated to a high degree, smelting and refining its mine output and using all or most of it in its own semi-fabricating and fabricating plants. Mineral processing and/or fabrication is usually carried on in the home country to service domestic markets. Functions such as purchasing are highly centralised, as is decision-making, and inputs utilised in mining are frequently obtained from within the multinational organisation. It is a model based largely on North American copper and aluminium companies which exploit minerals in South America and the Caribbean, usually for consumption in their home countries.

Some of the foreign mining companies encountered in this study displayed certain of these traits (particularly aluminium companies in Australia), but generally they have not corresponded to the model described above. They have typically not been vertically integrated beyond the refining stage, and have disposed of their output to independent smelters or fabricators. What processing they do carry out is usually performed in the host country. They exercise a substantial degree of autonomy in their purchasing policies, and usually do not purchase significant quantities of mining inputs from associated companies. Reflecting these characteristics, they generally do not engage in transfer pricing, they base investment decisions, including those relating to mineral processing, on

'normal' investment criteria (which include perceptions of risk in the host country), and they are
generally willing to utilise domestic sources of
goods and services.

The nature of the firms encountered reflected
the motivation of parent companies in undertaking
foreign investments and the industry structures
within which they operated. Most parent companies
were motivated by a desire to apply capital and
skills profitably in exploiting minerals for sale
outside their home countries, a fact reflected in
their organisational structures, which typically
consisted of a number of self-contained units, each
expected to generate a profit independently of the
others. Outside North America, the international
copper, lead and zinc industries within which many
of the companies operated are characterised by a
relatively low degree of concentration and of
vertical integration.

Second, 'objective' factors in the host country
play a very important part in determining the impact
of foreign mining investment. The existing level of
industrialisation influences the prospects for
backward linkage, proximity to markets for metal
goods and other factors of a similar kind influence
the prospects for forward linkage, the nature of the
national workforce determines the degree to which
diffusion of skills can occur, and existing patterns
of economic activity influence the impact of mine
infrastructure.

Third, political processes affect the impact of
foreign mining investment in a very wide variety of
ways, many of which are not recognised even by
authors such as Moran who accord a central role to
political factors. This last point brings us to the
question of host country-foreign investor relations.

It is apparent from the case studies that
shifts in bargaining power, declining uncertainty,
progress along learning curves and elite complicity
can explain certain aspects of host country-foreign
investor relations in particular situations. The
Papua New Guinea government's bargaining position in
relation to the Bougainville project was stronger in
1974 than the Australian Administration's had been
in 1967; BCL had made a major investment from which
it could not credibly threaten to withdraw as long
as it was allowed a 'reasonable' return, while the
profitability of the project had been amply demonstrated and the risks associated with the initial
investment had largely faded from view. The discovery of the Tynagh and Tara lead/zinc deposits in

Ireland in 1961 and 1970, respectively, represented important steps in a long-term decline in investor uncertainty as to conditions in the host country, a development which certainly improved the Irish government's bargaining position. The Papua New Guinea and Irish governments did move along relevant learning curves, and again this increased their potential bargaining powers. Irish and Australian authorities did fail to exercise their full bargaining powers, and in the Irish case this failure was associated with a 'pay-off' to foreign investors.

However, the usefulness of these concepts in explaining the outcome of foreign mining investments and in analysing host country-foreign investor relations is limited, partly because they do not encompass certain aspects of host country behaviour. For example, they ignore the fact that ignorance (which does not involve any element of complicity) can exert a major influence, and that a host country which perceives its interests as being compatible with those of the foreign investor may feel no need to overcome its ignorance. The concept of host country progress along a learning curve ignores the fact that a host country does not have to rely on its own resources in dealing with foreign mining companies, but can obtain expertise from abroad and thus avoid what may be the very substantial costs of _itself_ acquiring expertise (that is, moving along a learning curve). But their usefulness is also limited for more fundamental reasons.

The case studies have shown that the political factors which influence the host country-foreign investor relationship are more complex than the concepts of 'pay-off' or 'complicity' can describe. In addition, those processes interact with the two other factors mentioned above (the nature of the foreign investor and 'objective' factors in the host country) to determine the outcome of any particular investment. Thus the processes which result in a given outcome are much more complex than concepts such as 'shifts in bargaining power', 'decline in uncertainty', 'learning curve' or 'complicity' can encompass.

The case studies raised a number of points in relation to the political factors which influence the foreign investment process. First, factors unrelated to the host country-foreign investor relationship resulted in outcomes which favoured the foreign investor. In Ireland and Australia political norms, reflected in legal and consitutional constraints, limited the freedom of action of host country authorities. Political and bureaucratic actors unintentionally undermined the position of the agency responsible for monitoring the activities of foreign shareholders in Zambia, and created obstacles to the Irish government's acquisition of relevant expertise. Weaknesses in the general policy-making process lessened Zambia's ability to exercise increases in bargining power effectively.

This last point raises another issue. In general, the literature referred to in Chapter One is concerned with how and why changes in bargaining power occur and with whether or not they are exercised. Insufficient attention is paid to the manner in which favourable shifts in bargaining power are utilised, or to the fact that this can influence the host country-foreign investor relationship and the physical impact of foreign mining investment. For instance, both the Zambian and Papua New Guinea governments were in strong bargaining positions in the years after their countries received independence, and both exercised that power to increase revenue from foreign mining projects. Zambia did so without sufficient thought for the implications of collecting revenue in a particular way, while the opposite was the case in Papua New Guinea. The outcome had implications both for host country-foreign investor relations (conflictual in Zambia, consensual in Papua New Guinea), and for the efficiency of mining operations (wasteful use of mineral resources in Zambia by encouraging high-grading, efficient use of mineral resources in Papua New Guinea by lightening the tax burden on marginal operations). Political processes were crucial in determining the outcome in each case, but these had little to do with the host country-foreign investor relationship. Rather they reflected broader historical and political factors which affected the whole policy-making process.

There is also the question of whether a host country correctly assesses the extent to which favourable shifts in bargaining power have actually occurred and to which it has acquired the ability to duplicate the functions of the foreign investor. The Zambian government apparently overestimated the extent of both in 1969, and Zambia incurred substantial costs as a result; once again, weaknesses in the general framework or policy-making were largely responsible. Australia's Labor government misinterpreted the significance of changes in the international supply/demand situation for minerals, and consequently overestimated Australia's ability to obtain foreign capital and expertise without permitting direct foreign investment.

Political processes relate to the 'objective' factors discussed above in a number of ways. First, policy-makers may take account of their existence or may fail to do so, and this can modify the impact of such factors on the outcome of foreign mining investment. For example, after Independence the Zambian government was faced with certain 'objective' factors resulting from the wage structure of Zambia's economy. If these had been taken into account in formulating employment and localisation policies towards the mining industry, the adverse effect of foreign mining investment on domestic income distribution could have been mitigated. Second, 'objective' factors may themselves be modified by appropriate government action. For instance, the nature of the host country's workforce may be such that diffusion of skills cannot occur, but appropriate educational policies can favourably alter its character. The relevance of this possibility was very evident in the Zambian case.

The policy-making process might relate to the character of foreign mining companies in two ways. First, host country authorities may identify the type of company most likely to act in a manner compatible with their interests, and discriminate in favour of such companies in granting prospecting and exploration rights. For example, a country anxious to develop forward linkages may favour companies which do not already possess processing facilities elsewhere; a country seeking to develop backward linkages might favour companies not affiliated with producers of mining inputs. Second, as mentioned above, governments can encourage foreign mining companies to display certain characteristics (favourable or unfavourable), particularly by their taxation policies.

Given the range of factors which determines the impact of foreign mining investment and the extent of interaction between them, it is evident that such investment may produce a very wide variety of outcomes, each on balance favourable or unfavourable to the host country to varying degrees. Thus at least in theory any of the models described in Chapter One might accurately depict what occurs in individual cases. Indeed, variations on these models (some of which are mentioned below) may be required to describe particular instances of foreign investment, and the relevance of one or another might vary over time as, for example, changes occurred in 'objective' conditions or in government policies.

If the foreign investor provides capital and expertise which would not otherwise be available, if host country political processes are such as to allow a rational pursuit of policy aims, and if the foreign company involved displays characteristics consistent with achievement of these aims, either of the 'compatible' models might apply. If conditions governing mining operations were generally applicable and displayed sufficient flexibility, limited conflict of the type envisaged by Mikesell might be avoided. At the moment, Papua New Guinea approximates to this model.

If the host country does possess the ability to duplicate the foreign investor's functions, if political processes are favourable to pursuit of host country interests, and if the characteristics of the foreign investor are inimical to those interests foreign investors may, as Moran suggests, be 'pushed out' without substantial cost to the host country. If political processes are not favourable, foreign investment may continue, and an exploitative situation persist over extended periods of time. The case studies provided substantial evidence to support the points made in Chapter One in relation to Moran's prediction that exploitation will necessarily be temporary. Bureaucratic decision-makers are sometimes in a position to grant concessions to foreign investors through processes which are secret, and political actions which result in concessions are sometimes not associated in the public mind with the granting of those concessions. For these reasons, public pressure for termination of complicity does not always emerge, while the concessions granted by host country political actors are sometimes very difficult to reverse (as in the case of MINDECO's disbandment in Zambia).

If political processes are such that the ability of the host country to duplicate the functions of the foreign investor is not accurately assessed, he may be 'pushed out' and the host country may incur substantial costs as a result, as occurred in Zambia.

It is also possible that foreign investors may play a vital role in providing capital and expertise and may behave in a manner compatible with host country interests, but that host country authorities may fail to reap their full share of the benefits from mineral development because of ignorance on their part or because political processes affect their ability to do so. This would seem to have been generally the case in Australia.

A basic question remains. Is the nature of foreign mining companies, of 'objective' factors in the host country and of political processes which affect the impact of foreign mining investments such that, in a majority of cases, the impact of that investment will be favourable or unfavourable? The evidence provided by the case studies allows a number of points to be made.

First, in all the case study countries foreign investors played a crucial role in providing scarce risk capital and expertise; the Australian and Zambian studies indicate that it is difficult for a host country to obtain those resources without permitting direct foreign investment. Thus the economic opportunities created by mineral development would not have arisen or would have been substantially diminished if foreign investment had not occurred. Second, the extent to which host countries grasped those opportunities and acted to mitigate any adverse social and economic costs arising from mineral development depended almost entirely on their ability to acquire relevant information and to devise effective policy-making structures. Host country failure to optimise benefits and minimise costs was almost entirely due to historical, political and institutional factors which militated against effective policy-making, and which were generally unrelated to the host country-foreign investor relationship. Only in Ireland was evidence found that foreign investors had influenced the decision-making process so as to bring about outcomes favourable to themselves. Third, the companies encountered in this study generally displayed characteristics which were compatible with pursuit of host country interests, though a minority did display behaviour typical of the corporate model described by critics of foreign mining investment.

Multinational and international mining companies may all be dead in the long run but this study demonstrates that it will be in the interests of many host countries to keep at least some such companies alive for extended periods of time.

Notes

1. Interview with BCL's Chief Purchasing Officer, Panguna, 15 June 1978.
2. Between 1964 and 1970 Zambia's share of world copper production fell from 13.0 per cent to 10.7

per cent. <u>Zambia Mining Yearbook, 1972</u>, p. 19.
3. AAC, <u>Annual Report, 1976</u>, pp. 78-81.
4. By the 1960s RST's mining interests were almost
exclusively confined to Zambia, and it was consequently less likely than the widely-diversified AAC
to become aware of, or be in a position to grasp,
favourable investment opportunities.
5. Drysdall in Bostock and Harvey, <u>Economic Independence and Zambian Copper</u>, p.73.
6. Export receipt would be higher where the impact
of the larger volume of exports was sufficient to
outweigh that of the lower unit price. It should
also be remembered that the net foreign exchange
inflow from mining operations was in any case highly
sensitive to price fluctuations.

SELECT BIBLIOGRAPHY


1. Government Publications

Australia

(Printed by the Government Printer, Canberra)

Commonwealth of Australia <u>Commonwealth Parliament-</u>
<u>ary Debates</u>, 1970-1978
_____ Senate Select Committee on Foreign Ownership
and Control <u>Official Hansard Report</u>, June-Septem-
ber 1972
Bureau of Mineral Resources, Geology and Geophysics
<u>Australian Mineral Industry Annual Review</u>,
various issues
Fitzgerald,T.M. <u>The Contribution of the Mineral</u>
<u>Industry to Australian Welfare</u>, 1974
Industries Assistance Commission <u>Report, Petroleum</u>
<u>and Mining Industries</u>, May 1976

Ireland

(Printed by the Government Printer, Dublin)

Republic of Ireland <u>Dail Debates</u>, 1956-1977
Minister for Industry and Commerce, 'Report ... for
the six months ended June 30th, 1975, in accordance with Section 77 of the Minerals Development
Act 1940', 1975

Papua New Guinea

(Printed by the Government Printer, Port Moresby)

Territory of Papua New Guinea <u>House of Assembly</u>
<u>Debates</u>, 1964-1972
_____ <u>Programmes and Policies for the Economic</u>
<u>Development of Papua New Guinea</u>, 1968
Papua New Guinea <u>House of Assembly Debates</u>, 1972-
1975
_____ <u>National Parliament Debates</u>, 1975-1978

Zambia

(Printed by the Government Printer, Lusaka)

Select Bibliography

Republic of Zambia Zambia Hansard, 1964-1975
_____ Report of the Commission of Inquiry into the
    Mining Industry, 1966
_____ Towards Complete Independence. Address by His
    Excellency the President, Dr. K.D. Kaunda, to the
    UNIP National Council held at Matero Hall,
    Lusaka, 11 August 1969
_____ Zambia's Economic Revolution. Address by His
    Excellency The President, Dr.K.D. Kaunda, at Mul-
    ungushi, 19 April 1968
MINDECO Zambia Mining Year Book, 1970-1978
Ministry of Development Planning and National Guid-
    ance Second National Development Plan, 1972-1976,
    1971
Office of National Development and Planning First
    National Development Plan 1966-1970, 1966

2. Books and Monographs

Australian Mining Industry Council Minerals Industry
    Survey 1977 and 1978 (Canberra, 1979)
Baldwin,R.E. Economic Development and Export Growth:
    A study of Northern Rhodesia, 1920-1960 (Univer-
    sity of California Press, Berkeley, 1966)
Bedford,R. and Mamak,A. Compensating for Develop-
    ment: The Bougainville Case (Bougainville Special
    Publication No. 2, Bougainville Special Public-
    ations, Christchurch, 1977)
Bos,H.C., Sanders,S. and Secchi,C. Private Foreign
    Investment in Developing Countries: A Quantit-
    ative Study of the Macro-Economic Effects (D.
    Reidel Publishing Company, Dordrecht, 1974)
Bostock,M. and Harvey,C.(eds.) Economic Independence
    and Zambian Copper: A Case Study in Foreign
    Investment (Praeger, New York, 1972)
Elliott,C.(ed.) Constraints on the Economic Develop-
    ment of Zambia (Oxford University Press, Nairobi,
    1971)
Girvan,N. Copper in Chile: a Study in Conflict bet-
    ween Corporate and National Economy (Institute of
    Social and Economic Research, University of the
    West Indies, Jamaica, 1972)
International Bank for Reconstruction and Develop-
    ment The Economic Development of the Territory of
    Papua New Guinea (Johns Hopkins Press, Baltimore,
    1965)
Kapoor,A. and Grub,P.D.(eds.) The Multinational
    Enterprise in Transition: Selected Essays and
    Readings (Darwin Press, New Jersey, 1973)
Lanning,G. and Mueller,M. Africa Undermined: a his-
    tory of the mining companies and the underdevel-

opment of Africa (Penguin Books, London, 1979)

McKern,R.B. Multinational Enterprise and Natural Resources (McGraw-Hill Book Company, Sydney, 1976)

Martin,A. Minding Their Own Business: Zambia's Struggle Against Western Control (Hutchinson, London, 1972)

Mikesell,R.F.(ed.) Foreign Investment in the Petroleum and Minerals Industry: Case Studies of Investor-Host Country Relations (Johns Hopkins Press, Baltimore, 1971)

_____ Foreign Investment in Copper Mining: Case Studies of Mines in Peru and Papua New Guinea (Johns Hopkins Press, Baltimore, 1975)

Moran,T.H. Multinational Corporations and the Politics of Dependence: Copper in Chile (Princeton University Press, Princeton, 1974)

Nziramasanga,M. 'The Copper Export Sector and the Zambian Economy', unpublished PhD thesis, Stanford University, 1973

O'Kelly,M. 'Legislation and the Irish Mining Industry', unpublished MBA dissertation, University College Dublin, 1978

Overseas Development Group (University of East Anglia) A Report on Development Strategies for Papua New Guinea (Port Moresby, 1973)

Raggatt,H.G. Mountains of Ore (Lansdowne Press, Melbourne, 1968).

Sklar,R. Corporate Power in an African State: The Political Impact of Multinational Mining Companies in Zambia (California University Press, Berkeley, 1975)

Stevenson,G. Mineral Resources and Australian Federalism (Centre for Research on Federal Financial Relations Research Monograph No 17, The Australian National University, Canberra, 1976)

United Nations Negotiating and Drafting of Mining Development Agreements: an interregional workshop, Buenos Aires, 1973 (Mining Journal Books, London, 1976)

Vernon,R. Sovereignty at Bay: The Multinational Spread of US Enterprises (Basic Books, New York, 1971)

World Bank Papua New Guinea: Its Economic Situation and Prospects for Development (Washington, D.C., 1978)

3. Articles, Papers and Pamphlets

Australian Mining Industry Council Mining Taxation - A Review (Canberra, June 1974)

_____ Mining Taxation and Australian Welfare
     (Canberra, June 1974)
Brundenius,C. 'The Anatomy of Imperialism: The Case
     of the Multinational Mining Corporations in
     Peru', Journal of Peace Research, 3, 1972
Evans,P.E. 'National Autonomy and Economic Develop-
     ment: Critical Perspectives on Multinational
     Corporations in Poor Countries', International
     Organization, Vol. XXV, No. 3 (Summer 1971)
Garnaut,R. and Clunies Ross,A. 'Uncertainty, Risk
     Aversion and the Taxing of Natural Resource Pro-
     jects', Economic Journal, Vol. 85 (June 1975)
Girvan,N. 'Multinational Corporations and Dependent-
     Underdevelopment in Mineral Export Economies,
     Social and Economic Studies, Vol. 19, No. 4
     (December 1970)
Hymer,S. 'The Efficiency (Contradictions) of Multi-
     national Corporations', American Economic
     Review: Papers and Proceedings, May 1970
Litvak,I.A. and Maule,C.J. 'Forced Divestment in the
     Caribbean', International Journal, Vol. XXXII,
     No. 3 (Summer 1977)
Moran,T.H. 'The Theory of International Exploitation
     in Large Natural Resource Investment', in
     Rosen,S.J. and Kurth,J.R., Testing Theories of
     Economic Imperialism (Lexington Books, Lexington,
     1974)
Ndulo,M. 'The Requirement for Domestic Participation
     in New Mining Ventures in Zambia', African Social
     Research, 25 (June 1978)
Singer,H.W. 'U.S. Investment in Underdeveloped
     Areas: The Distribution of Gains Between Inves-
     ting and Borrowing Countries', American Economic
     Review, Vol. XL (May 1980)
Treadgold,M. The Regional Economy of Bougainville:
     Growth and Structural Change (Development Studies
     Centre Occasional Paper No. 10, The Australian
     National University, 1978)
Wells,L.T.Jr. 'The Multinational Business Enter-
     prise: What Kind of International Organization?',
     International Organization, Vol. XXV, No. 3
     (Summer 1971)

294

Index

Index

Zambia 145, 147
wage payments by foreign
  mining companies 58,
  132-4, 258
Wells, L.T. 246
Western Australia 163,
  190, 196
Western Mining Corporat-
  ion (WWC) 196, 204-5
Whitlam, G. 174-6, 241
World Bank 115, 242

Zambia 96-161 passim,
  273-88 passim. See
  also Northern Rhodesia
Zambianisation 107-8,
  114, 128, 130-1,
  140-3, 144, 147-8
ZIMCO 109, 111, 115,
  126-7,
zinc smelter, Ireland
  59-60, 69n53, 82,
  83, 89

For Product Safety Concerns and Information please contact our EU
representative GPSR@taylorandfrancis.com
Taylor & Francis Verlag GmbH, Kaufingerstraße 24, 80331 München, Germany

* 9 7 8 1 1 3 8 0 8 3 7 1 4 *